ELECTRICAL, MAGNETIC, AND VISUAL METHODS OF TESTING MATERIALS

Jack Blitz, M.Sc., F.Inst.P., M.NDT.S.
Brunel University, London

William G. King, A.I.M.
Rolls Royce Ltd., Small Engine Division

and

Donald G. Rogers, M.NDT.S., A.M.I.E.I.
Rolls Royce Ltd., Small Engine Division

LONDON

BUTTERWORTHS

ENGLAND: BUTTERWORTH & CO. (PUBLISHERS) LTD.
LONDON: 88 Kingsway, W.C.2

AUSTRALIA: BUTTERWORTH & CO. (AUSTRALIA) LTD.
SYDNEY: 20 Loftus Street
MELBOURNE: 343 Little Collins Street
BRISBANE: 240 Queen Street

CANADA: BUTTERWORTH & CO. (CANADA) LTD.
TORONTO: 14 Curity Avenue, 374

NEW ZEALAND: BUTTERWORTH & CO. (NEW ZEALAND) LTD.
WELLINGTON: 49/51 Ballance Street
AUCKLAND: 35 High Street

SOUTH AFRICA: BUTTERWORTH & CO. (SOUTH AFRICA) LTD.
DURBAN: 33/35 Beach Grove

Suggested U.D.C. number: 620·179·1
Standard Book Number: 408 18350 0

Printed in Great Britain by
R. J. Acford, Ltd., Chichester, Sussex

PREFACE

In 1960 the book *Techniques of Non-Destructive Testing*, edited by C. A. Hogarth and myself and published by Butterworths, made its appearance. It contained short and concise contributions from different authors, including W. G. King and J. C. Rockley, on the various non-destructive testing techniques such as x-ray, isotope, ultrasonic, eddy current, magnetic particle, and penetrant methods. That book is now out of print and the preparation of a second edition has been made difficult because some of the authors are no longer available.

In the meantime, however, Butterworths have published Rockley's *An Introduction to Industrial Radiology*, dealing with x-ray and isotope methods, and the English translation of *Ultrasonic Methods of Testing Materials* by L. Filipczynski, Z. Pawlowski, and J. Wehr, which I edited. The publication of *Electrical, Magnetic, and Visual Methods of Testing Materials* thus completes a three-volume series covering the conventional methods of non-destructive testing in a more comprehensive manner than the original single volume.

This present work concentrates mainly on eddy current, magnetic particle, and penetrant methods of testing but it also deals with a number of other electrical, magnetic, and optical techniques, in rather less detail. Brief descriptions are given of one or two of the latest developments in non-destructive testing, including holography and the use of microwaves.

Advantage has been taken of the wealth of experience built up over many years by Messrs. King and Rogers in magnetic and penetrant methods by providing detailed descriptions of the correct procedure for these techniques and pointing out the many pitfalls which the inexperienced and unwary operators are likely to meet. The application of these methods must be regarded as more of an art in comparison with those of eddy currents, for which the operator can rely on the use of modern electronic equipment. Even so, details of the procedure with the Magnatest Q, a typical eddy current device, for the testing of small components are provided. These are given in an Appendix to this book in order to preserve the balance in Chapter 2. It had been hoped to give fuller details of the procedure with other eddy current methods, in the light of practical experience gained by the authors, but this was not possible in view of internal security regulations.

PREFACE

The opinions expressed in Chapters 3, 4 and 6, in the Appendix, and in parts of Chapter 2 are not necessarily shared by the authors' (W.G.K. and D.G.R.) present employers, Messrs. Rolls-Royce Ltd. (Small Engine Division), whose permission to use many of the illustrations is gratefully acknowledged.

The authors also wish to thank the undermentioned for their kind assistance:

Professor C. A. Hogarth, Dr. P. A. Feltham and Dr. G. F. Lewin of Brunel University; P. W. Allen and Co. Ltd.; Ardrox Ltd.; Automation Industries U.K. (formerly Budd Instruments U.K.); British Aluminium Co. Ltd.; C.N.S. Instruments Ltd.; E.S.A.B. Ltd.; Engelhard Hanovia Lamps Ltd.; Evershed and Vignoles Ltd.; Fel-Electric Ltd.; Forster Instruments Ltd.; G.E.C.-A.E.I. (Electronics) Ltd.; Magnaflux Corporation; Microwave Instruments Ltd.; The Nondestructive Testing Centre, Harwell; Pantak Ltd.; Radalloyd Ltd.; Stevic Engineering Ltd.; Ultrasonoscope Co. (London) Ltd.; Vitosonics Ltd.

J. BLITZ

Brunel University

CONTENTS

CONTENTS

NOTES ON UNITS

To those unfamiliar with the S.I. units used in this book, the following remarks may prove useful.

Mass, length, and time are expressed in the metric system in terms of the metre (m), kilogramme (kg) and second (s). The unit of frequency is the Hertz (Hz), which is equivalent to the cycle per second (c/s). Multiples and sub-multiples of these units are related to one another by integral powers of 10^3 using the following prefixes:

$$
\begin{aligned}
\text{pico (p)} \ \ &= \ 10^{-12} \\
\text{nano (n)} \ \ &= \ 10^{-9} \\
\text{micro } (\mu) \ \ &= \ 10^{-6} \\
\text{milli (m)} \ \ &= \ 10^{-3} \\
\text{kilo (k)} \ \ &= \ 10^{3} \\
\text{mega (M)} \ \ &= \ 10^{6} \\
\text{giga (G)} \ \ &= \ 10^{9}
\end{aligned}
$$

For example:

$$
\begin{aligned}
1 \text{ nm (1 nanometre)} \ \ &= \ 10^{-9}\text{m} \\
1 \mu\text{s (1 microsecond)} \ \ &= \ 10^{-6}\text{s} \\
1 \text{ GHz (1 gigaHertz)} \ \ &= \ 10^{9}\text{Hz (i.e. } 10^{9}\text{c/s)}
\end{aligned}
$$

ELECTRICAL AND MAGNETIC METHODS OF TESTING MATERIALS—GENERAL

1.1. GENERAL CONSIDERATIONS

THE MECHANICAL behaviour of a substance can usually be related in some way to its electrical and magnetic properties, i.e. electrical conductivity, magnetic permeability, and dielectric behaviour. For example, when a piece of steel is work-hardened, there are changes in both its electrical resistance and magnetic permeability. Measurements of one or both of these changes can provide an indication of the degree of work-hardening.

Materials may be classified in accordance with their electrical conductivities as follows:

(1) *Good conductors.* They include all metals and certain non-metals such as carbon. The values of their electrical resistivities (see Section 1.2) are of the order of 10^{-8} to 10^{-7} ohm m (10^{-6} to 10^{-5} ohm cm) at room temperature (see *Table 1.1*).

(2) *Semi-conductors.* These materials have electrical resistivities of the order of 10^{-2} ohm m (1 ohm cm) at room temperature. The best known examples are germanium, silicon, and cadmium sulphide.

(3) *Poor conductors.* Poor conductors, also known as insulators, have very high electrical resistivities at room temperature, often higher than 10^{12} ohm m (10^{14} ohm cm).

Materials can also be classified with respect to their magnetic properties. Substances which can be strongly magnetized include ferromagnetic materials, such as nickel and certain steels, and ferrimagnetic materials, also known as ferrites or *ferroxcubes*. The latter are ceramic oxides which are non-metallic and usually semi-conductors. Ferromagnetic substances, on the other hand, are always good conductors of electricity.

The extent to which a given material can be magnetized is characterized by its *magnetic permeability* (see Section 1.3); ferromagnetic materials and ferrites can possess very high values of relative permeability, often of the order of hundreds or even thousands. Materials which cannot be magnetized to any noticeable extent have relative permeabilities very close to unity and may be described as being either *paramagnetic* or *diamagnetic*. These terms, which are explained in most textbooks on electricity and magnetism (see, for example, Scott[1]), are of considerable importance in the study of the fundamental properties of materials but are of little interest from the point of view of the routine testing of the mechanical properties of manufactured components.

Another method of classifying materials is in respect of their dielectric behaviour. For a given substance, this depends on the *dielectric constant* (or *relative electrical permittivity*) and the *dielectric loss* (or *dissipation factor*) (see Section 1.4). In practice these terms apply only to what are called dielectric materials, in which electric fields can be produced. The properties of dielectric materials, which may be either poor conductors or semiconductors, can be investigated by the uses of capacitor, microwave, and electrified particle techniques.

1.2. THE PROPERTIES OF ELECTRICAL CONDUCTORS

The properties of an electrically conducting material can often be assessed from the measurement of its *electrical resistance R*, defined as the ratio of the *potential difference* (or *voltage*) V between the extremities of the conductor to the *current I* flowing through it. If V is measured in volts and I in amperes, R is expressed in ohms. The value of R depends not only on the physical properties of the material but also on its shape and size. It is therefore desirable to define a more basic quantity, namely the electrical *resistivity* (or *specific resistance*) ρ, which is independent of the geometrical characteristics of any particular sample. The reciprocal of the resistivity is the *conductivity* σ. For a conductor of length l and uniform cross-section of area A, providing that its material is homogeneous, we have

$$\rho = 1/\sigma = RA/l \qquad (1.1)$$

ρ is usually expressed in ohm-m or ohm-cm, depending on the system of units employed and σ in mho/m or mho/cm. Values of ρ and σ for a number of materials are given in *Table 1.1*.

A unit of conductivity which has found favour with engineers is the m/ohm mm², which is equivalent to 10^6 mho/m. It has the advantage that the numerical values of the conductivities of the more common metals have a convenient order of magnitude. For example, the approximate conductivities of copper and aluminium

1.2. THE PROPERTIES OF ELECTRICAL CONDUCTORS

Table 1.1. Approximate Values of Electrical Resistivities and Conductivities a Room Temperature for Some Materials in Common Use

(The exact values depend on the structures, conditions and impurity contents of the materials)

Material	Resistivity ρ ohm-metre*	Conductivity σ mho/metre†
Aluminium	$2\cdot5 \times 10^{-8}$	$4\cdot0 \times 10^{7}$
Copper	$1\cdot7 \times 10^{-8}$	$6\cdot0 \times 10^{7}$
Graphite	$2\cdot0 \times 10^{-8}$	$5\cdot0 \times 10^{7}$
Iron	$8\cdot9 \times 10^{-8}$	$1\cdot2 \times 10^{7}$
Nickel	$6\cdot1 \times 10^{-8}$	$1\cdot6 \times 10^{7}$
Platinum	$9\cdot8 \times 10^{-8}$	$1\cdot0 \times 10^{7}$
Silver	$1\cdot5 \times 10^{-8}$	$6\cdot7 \times 10^{7}$
Tin	$11\cdot5 \times 10^{-8}$	$0\cdot9 \times 10^{7}$
Zinc	$5\cdot1 \times 10^{-8}$	$2\cdot0 \times 10^{7}$
Germanium $\Big\}$ Silicon	10^{-5} to 10^{-1}	10 to 10^{5}
Ebonite	10^{14}	10^{-14}
Glass	10^{6} to 10^{12}	10^{-12} to 10^{-6}
Mica	10^{11} to 10^{13}	10^{-13} to 10^{-11}
Nylon	10^{11} to 10^{13}	10^{-13} to 10^{-11}
Perspex	10^{13}	10^{-13}
PVC	10^{10} to 10^{14}	10^{-14} to 10^{-10}
Porcelain	10^{11} to 10^{13}	10^{-13} to 10^{-11}
Rubber	10^{13} to 10^{16}	10^{-16} to 10^{-13}

*For values in ohm-cm, multiply by 10^{2}
†$\begin{cases} \text{For values in mho/cm, divide by } 10^{2} \\ \text{For values in m/ohm. mm}^{2} \text{ divide by } 10^{6} \\ \text{For values in \%IACS, multiply by } 1\cdot7421 \times 10^{-6} \end{cases}$

are 60 and 40 m/ohm. mm², respectively. Another unit of conductivity is the International Annealed Copper Standard (IACS). The conductivity of 99·999 per cent pure annealed copper is said to be 100 per cent IACS and 1·7421 per cent IACS is equal to 1 m/ohm. mm². The approximate value of the conductivity of aluminium is then 67 per cent IACS.

For good conductors, the resistivity usually increases with temperature, whereas for insulators and semiconductors, there is generally a decrease in resistivity with increase in temperature. Thus the measurement of the resistance of a sample of material can be related to either the resistivity, which is characterized by the physical properties of the material, or the shape and size of the sample, provided that one of the two parameters is either known or can be kept constant. For example, if several samples have the same shape and size, then a given property of the material, such as work-hardening or degree of heat-treatment, can be investigated. *Figure 1.1* shows the relation between the resistivity of a silver wire subjected

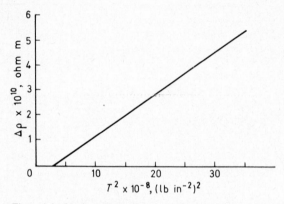

Figure 1.1. Variation $\Delta\rho$ of the resistivity of a silver wire with the square of the tensile flow stress T at 28°C (after Kovacs, Nagy and Feltham[14])

to plastic deformation and the tensile flow stress. Alternatively, if it is known that all samples of a given batch are homogeneous and have identical physical properties, resistance measurements can be used to determine size variations, such as thickness changes in sheets of metal.

The various methods of measuring resistance are described in most electricity textbooks (see, for example, Scott[1]). Precision measurements involve the use of a Wheatstone bridge. However, a straightforward method which is sufficiently accurate for many purposes consists of measuring voltage and current simultaneously and then dividing the first quantity by the second quantity.

An important application of resistance measurements is a device used by Valdes[2] for the assessment of the purity of germanium samples. It can also be employed for investigating local variations of resistivity caused, for example, by lack of homogeneity resulting from previous metallurgical treatment or by the presence of mechanical defects such as cracks, inclusions, and porosity. The device, illustrated in *Figure 1.2a*, consists of four spring-loaded probes with their ends always in contact with the material under test. A steady low direct current enters the material from the probe A and leaves it through the probe B. The potential difference appearing between C and D is measured; its value depends on the electrical resistance between these points. Initially the device must be calibrated with measurements made on a standard sample of known resistivity. The probe is then moved over the surface of the test sample; any variations in the observed potential difference would indicate a lack of homogeneity caused, perhaps, by structural variations or by the existence of a defect. *Figure 1.2b* indicates the paths of electric currents which are deflected by a crack. Along each line the current has a constant value.

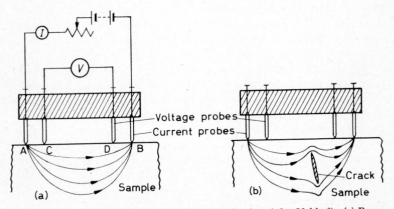

Figure 1.2. Four-probe device for resistivity investigations (after Valdes[2]). (*a*) Regular lines of current flow in the absence of a defect. (*b*) Distorted lines of current flow due to the presence of a crack

Valdes's method is generally suitable only for materials of high resistivity. Where the sample is a good conductor, small contact resistances of uncertain values, which can well have the same order of magnitude of the resistance of the sample itself, may present difficulties. In this case, the eddy current methods described in Chapters 2 and 3 are preferred because no actual contact with the test material takes place and, furthermore, it is possible to investigate simultaneously those phenomena which give rise to changes both in resistivity and dimensions. If, however, the surface of the sample has been carefully prepared, the four-probe method can be used for the measurement of the depth of cracks in good conductors, such as steel and aluminium. A modification of the four-probe device, developed in Germany by the Karl Deutsch organization and marketed in Great Britain by Vitosonics Ltd., can be used to measure crack depths of up to 100 mm to an accuracy of better than 10 per cent in suitably prepared samples.

An instrument which operates on the principle of resistance measurement, although not of the actual material under test, is the *strain gauge*. This is a conductor placed in adhesive contact with the test sample to which variations in dimensions and changes in position of the sample are transmitted. Very small dimensional changes can be measured to high degrees of precision because the changes in the resistance of the gauge with deformation are relatively high. For example, the strain gauge can be used for investigating rates of creep in materials under stress and dimensional variations of as little as 10^{-7} inch (4×10^{-6} mm) per day can be measured without difficulty. A number of different types of strain gauge are in common use; for details see Perry and Lissner[3].

De Meester, Deknock, and Verstappen[4] have described the application by Bastogne of a strain gauge for determining thicknesses of very narrow gaps, such as those between fuel element plates in nuclear reactors. The apparatus used (see *Figure 1.3*) consists of a curved metal blade with three small ruby wheels at one side and one ruby wheel at the centre of the opposite side. The strain gauge is attached to the outside of the blade. Variations in the curvature

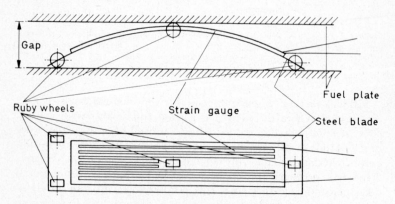

Figure 1.3. Bastogne's strain gauge device for measuring thicknesses of gaps between fuel element plates in nuclear reactors (de Meester, Deknock, and Verstappen[4]) (*By courtesy of the Society for Nondestructive Testing Inc.*)

of the blade, resulting from thickness changes in the gap, are indicated by the gauge.

1.3. THE MAGNETIC PROPERTIES OF MATERIALS

1.3.1. GENERAL CONSIDERATIONS

Ferromagnetic and ferrimagnetic materials can be tested in a non-destructive manner by measuring their magnetic permeabilities, either directly or indirectly. When a sample of one of these materials is placed in a magnetic field it becomes magnetized to an intensity depending on what is known as the *magnetic flux density* (sometimes called magnetic induction) B. The relationship between the intensity H of the magnetic field and the flux density is given by the following expression

$$B = \mu H = \mu_r \mu_o H \tag{1.2}$$

In the S.I. system of units, B is expressed in webers/m² and H in amperes/m (A/m) (or ampere turns/m). μ is called the *magnetic permeability* and is given in henrys/m (H/m). It is common practice to describe this quantity in terms of the *relative permeability* μ_r, which is the ratio of μ for the material to the permeability μ_o for free space. The value of μ_o is $4\pi \times 10^{-7}$ H/m. μ_r is a dimensionless number equal to unity for free space and very near unity for non-ferromagnetic materials. As stated earlier, the values of μ_r for ferro- and ferrimagnetic materials vary, depending on the nature of the material and the intensity of the applied magnetic field, and can be as high as several thousand.

The magnetic properties of a substance can be investigated by a number of methods, including the following:

(a) A.C. flux measurements whereby the inductance of a coil, placed near the test sample, is determined.

(b) D.C. flux measurements, usually involving the investigation of the Hall effect in a semiconductor element placed near the test sample.

(c) Observing the mechanical force of attraction between the test sample and a magnet.

1.3.2. A.C. FLUX MEASUREMENTS

If a current flows through a coil, a magnetic field acting in a direction at right-angles to the plane of the coil is observed and a magnetic flux will thus pass through any material in the vicinity of the coil. For an alternating current, both field and flux vary

7

periodically. The value of the flux density B at a given point in a material depends on the current, the position of the point in relation to the coil, and the self-inductance L of the coil. For a system in which the coil either surrounds or is placed near a test sample of permeability μ, provided that there are no other materials in the vicinity, the value in henrys of L for the coil is given by

$$L = Kn^2A\mu \qquad (1.3)$$

where n is the number of turns per metre length of the coil, A the mean area in square metres of cross-section of the coil, and K a dimensionless number, the value of which depends on the sizes, shapes, and relative positions of both coil and sample. For a given system, μ for the material can be determined by measuring L.

L is often measured with an inductance bridge. *Figure 1.4* illustrates the circuit diagram of a simple form of this type of a.c. bridge.

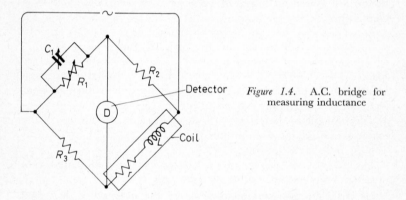

Figure 1.4. A.C. bridge for measuring inductance

R_2 and R_3 are fixed resistances of suitable known values and R_1 is a resistance of known value which can be varied in fixed stages. C_1 is a calibrated variable capacitance and r represents the value, usually unknown, of the resistance of the coil L. The source is an oscillator operating at a frequency of, say, 1,000 Hz and the detector can be an oscilloscope, a valve-voltmeter, a thermo-galvanometer, or a pair of headphones. C_1 and R_1 are varied alternately until the signal indicated by the detector is at a minimum level, for which the following relationships apply:

$$L = R_2R_3C_1 \qquad (1.4)$$

$$r = R_2R_3/R_1 \qquad (1.5)$$

The value of L is obtained from equation 1.4. Fuller details of inductance bridges are given by Hague[5].

The inductance of a coil can also be determined by measuring the voltage across its ends when an alternating current of known value flows through it. Provided that the resistive component of the impedance of the coil is small, we have that

$$\omega L = V_{rms}/i_{rms} \tag{1.6}$$

where ω is the angular frequency and V_{rms} and i_{rms} the root mean square values of voltage and current, respectively.

The foregoing discussion has not taken into account the induction of eddy currents in the sample by the alternating current flowing through the coil. It is shown in Section 2.2 that, as a result of this phenomenon, the flux density B depends not only on the magnetic permeability μ of the sample but also on the electrical conductivity σ. There is consequently a frequency dependent component of resistance which must be added to the ohmic resistance of the coil. By resolving the two components of the impedance of the coil, both the permeability and conductivity of the material under test can be determined. If, however, one is interested only in permeability, the inductance should be measured at a very low frequency, say 10 Hz, at which eddy current effects are minimum. Further details of eddy current measurements for ferromagnetic materials are given in Sections 2.2.3, 2.4.8 and the Appendix.

Eddy current effects can be practically eliminated by increasing the flux density through the coil by means of a permanent magnet. This principle is used by Förster in the design of his Precision Sheet Thickness Gauge (see *Figure 1.5*) for measuring thicknesses of sheets of ferromagnetic metal. A single probe houses both coil and magnet. An accuracy of approximately from 2 to 3 per cent over the range 0 to 0·6 mm is claimed. Thicknesses of up to 2·5 mm can be measured by this method.

De Meester, Deknock and Verstappen[4] describe an indirect application of inductance measurements to thickness gauging. The device, consisting of a spring-loaded ferrite core inside a coil, is used to determine thicknesses of gaps which might be inaccessible to other types of thickness gauge. The position of the core in relation to the coil, as determined by the gap thickness, governs the amount of magnetic flux passing through the coil and hence the inductance of the latter. Uncertainties may be introduced where the walls of the gap are made of ferromagnetic material.

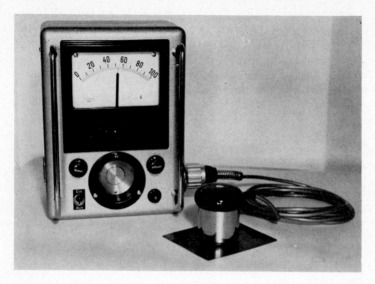

Figure 1.5. Precision sheet thickness gauge, Type 2.203 (*By courtesy of Forster Instruments Ltd.*)

1.3.3. D.C. FLUX MEASUREMENTS

Difficulties arising from the induction of eddy currents are avoided if d.c. measurements are made. Several different types of instrument, including magnetometers and fluxmeters, can be used but they may not be adaptable for use with the types of probe coils suitable for the conventional non-destructive testing applications. By far the most suitable d.c. method for measuring flux density and permeability makes use of the Hall effect.

The Hall effect is observed when a direct current i through a conductor flows at right-angles to a steady magnetic field H (see *Figure 1.6*). A direct voltage is then generated across the edges of the conductor in a direction mutually perpendicular to the current and the field. If V is the voltage, B the flux density arising from the field, J the current density and W the width of the conductor, we have

$$V = WRJB \qquad (1.7)$$

where R is defined as the Hall coefficient, which is constant for a given conductor. Thus, for a constant current, V varies directly with B. Measurements are more sensitive when using semi-conductor elements, for which the value of the Hall coefficient R is considerably higher than for metals.

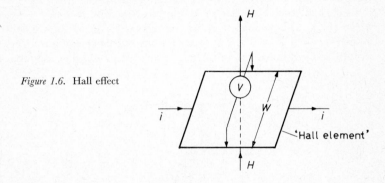

Figure 1.6. Hall effect

Mix[6] describes a device which applies the Hall effect to the measurement of thicknesses of platings of ferromagnetic materials such as nickel and steel on non-ferromagnetic bases. The instrument, illustrated in *Figure 1.7*, contains a Hall element (or 'Hall-Pak')

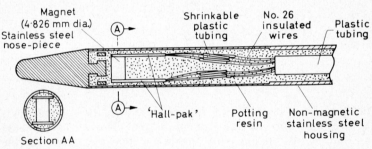

Figure 1.7. Hall probe for measuring the thickness of nickel plating (Mix[6]) (*By courtesy of the Society for Nondestructive Testing Inc.*)

consisting of two plates of the semiconductor indium arsenide and a strong permanent magnet which provides a steady magnetic field. It is fitted with a stainless steel nose-piece. When the latter is placed in contact with the surface of the ferromagnetic test material, there is an increase in the flux density across the Hall element and thus a change in the measured voltage. The device, which is calibrated for known thicknesses of plating, provides an accuracy of $\pm 1\mu m$.

Förster (see McMaster[7]) has applied the Hall effect to the measurement of crack depths in ferromagnetic conductors. A direct current is passed through the material in the same way as with the four-probe method of Valdes described in Section 1.2 (see *Figure 1.2*). The voltage probes are replaced by a single probe consisting of a Hall element placed midway between the current probes and just above the test surface (see *Figure 1.8*). The magnetic

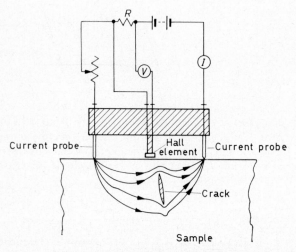

Figure 1.8. Hall probe for measuring crack depths in ferromagnetic materials (after Förster, see McMaster[7])

field resulting from the direct current through the specimen acts at right-angles to the plane of the paper. This method has the advantage over the four-probe method in that there is no trouble from contact resistances.

1.3.4. MAGNETIC FORCES OF ATTRACTION

The force of attraction between a magnet and an unmagnetized ferromagnetic material is a well-known phenomenon. For a given magnet, the magnitude of this force is dependent on the value of the magnetic permeability and size of the unmagnetized material and on the distance between the two bodies.

This phenomenon has been applied to the design of a pocket-sized instrument used for measuring the thickness of a thin layer of non-ferromagnetic material (e.g. paint, plastic, copper plating) on a ferromagnetic surface (see *Figure 1.9*). It consists essentially of a pencil-like tube, open at its lower end, which contains a magnet suspended from a spring fixed to the upper end. When not in use, the bottom of the magnet is level with the lower end of the tube.

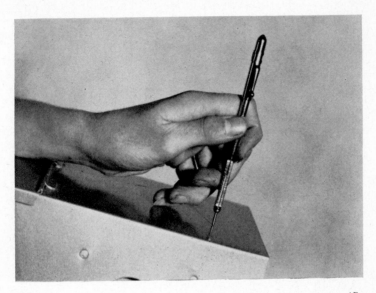

Figure 1.9. B.S.A. Tinsley gauge for paint thickness measurements (*By courtesy of Evershed and Vignoles Ltd.*)

The device is placed vertically on the surface under test and slowly raised until the tension in the spring is just sufficient to overcome the magnetic force of attraction. The length of the magnet exposed by this action is dependent on the thickness of coating. The device must be calibrated for coatings of known thicknesses. By this method, one can measure thicknesses of up to 0·5 mm with an accuracy of about 10 per cent.

An indirect application to non-destructive testing of magnetic forces of attraction is the magnetic particle method discussed in some detail in Chapters 3 and 4.

1.3.5. MAGNETIC HYSTERESIS

It has already been pointed out that the magnetic permeability of a ferromagnetic material is dependent on the intensity of the applied magnetic field; thus the relationship between B and H, as given in equation 1.2, is not a linear one.

Consider an unmagnetized bar of ferromagnetic material placed inside a coil connected to a d.c. source. When the circuit is switched on, a magnetic field H, proportional to the current i in the coil, acts along the direction of the axis and a flux density B is induced in the material. If H is slowly increased from zero by increasing i, an increase in B is observed as indicated by the *magnetization curve* OAB (see *Figure 1.10*). The gradient of this curve is equal to the *differential permeability* μ_d, defined as follows

$$\mu_d = dB/dH \qquad (1.8)$$

The value of μ_d will, in general, not be the same as that of the absolute permeability μ defined in equation 1.2. Beyond point A the slope of the curve is constant and the value of μ_d becomes equal

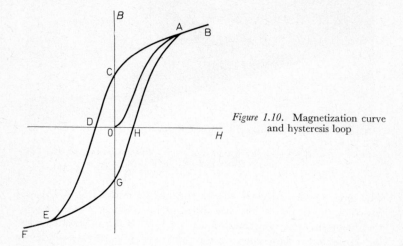

Figure 1.10. Magnetization curve and hysteresis loop

to the permeability μ_o of free space. The specimen then becomes magnetically saturated and no further magnetization is possible no matter by how much the field H is increased. This phenomenon is used to advantage for the eddy current testing of ferromagnetic materials if one wishes to eliminate all magnetic effects (see Section 2.2.3). In this case, the incremental value dH of the field must not be large enough to destroy the saturation during the negative part of the cycle.

On decreasing H, the curve does not return along its original path but, instead, follows the line BAC. At the point C, for which the field H is zero, the flux density B has a value B_r called the *retentivity* or *remanence*. When the direction of H is then reversed and its value increased from zero, the curve continues along the path CDEF. At D, where B is zero, the field H has a value H_c called the *coercive force*. At E saturation again occurs and, beyond this point, μ_d becomes equal to μ_0. On decreasing H, the curve follows the path FEG and, on again reversing the direction of H and increasing its value, the curve follows the line GHAB. It is seen that the curves contain a closed loop ACDEGHA; this is called a *hysteresis loop*, the shape of which is dependent on the physical properties of the material, e.g. structure, hardness, and anisotropy. The same loop will be formed for a given material when H is varied cyclically, provided that saturation level is reached during each half-cycle.

The hysteresis method of testing ferromagnetic materials has been developed by Förster to a considerable extent. One of his instruments, called the *Ferrograph* (see McMaster[8]), consists of two coils, one for magnetizing the specimen and the other for detecting the flux density B. H is varied sinusoidally by passing an alternating current through the magnetizing coil; B will thus vary in a similar manner with respect to time. The detecting and magnetizing coils are connected, via separate integrating circuits, to the Y- and X-plates, respectively, of a cathode-ray oscilloscope. The voltage input to the Y-plates is proportional to the flux density B and that to the X-plates is proportional to the field H. A stationary trace of the hysteresis loop is thus observed on the oscilloscope screen. Allowance must be made for the induction of eddy currents in the sample.

Another instrument devised by Förster for hysteresis testing is the *Magnatest Q*, which is essentially an eddy current device, and descriptions of its principles and applications are given in Section 2.4.8 and the Appendix.

Eddy current induction can be minimized by working at a very low frequency (say 10 Hz). This effect can be avoided altogether with d.c. measurements for which B may be measured with a Hall element probe.

A fuller account of hysteresis methods of non-destructive testing is given by McCaig[9].

1.4. THE PROPERTIES OF DIELECTRIC MATERIALS

1.4.1. GENERAL CONSIDERATIONS

The essential property of a dielectric is the absence of free electric charges which would otherwise give rise to electric currents. Thus it can contain an electric field. If, in a dielectric material, a steady

15

potential difference of V volts is maintained between two points A and B at a distance d from one another, an electric field is set up between those points. The intensity of the field at different positions between A and B depends on the shape and size of the material, on variations of the dielectric properties in the substance, and on the presence of neighbouring electrical conductors. However, the *average* intensity of the field is equal to V/d volts/m.

1.4.2. TESTING WITH A CAPACITOR

A direct method of measuring the dielectric properties of a material is to sandwich it between two metal plates so as to form a capacitor (condenser). The electrical properties of a capacitor depend on the value of its *capacitance* C, defined as follows

$$C = Q/V \qquad (1.9)$$

where Q represents the amount of charge stored (i.e. $+ Q$ on one plate and $- Q$ on the other) and V the potential difference across the plates. If Q is measured in coulombs (i.e. ampere-seconds) and V in volts, C is expressed in farads. A capacitance of one farad is very large and is met with only in high-voltage applications. Normally one deals with capacitances of the orders of microfarads (1μF $= 10^{-6}$ F) and picofarads (1 pF $= 10^{-12}$ F).

The capacitance C of a capacitor consisting of two parallel plates, each of area A square metres, distance d metres from one another and separated by a dielectric material, is given by the expression

$$C = \epsilon A/d \qquad (1.10)$$

where ϵ represents the *electrical permittivity* of the material. ϵ, which is expressed in farads per metre, is analogous to the magnetic permeability μ describing the magnetic properties of a material (see equation 1.2). The *relative permittivity*, or *dielectric constant*, k is a dimensionless number defined as follows

$$k = \epsilon/\epsilon_0 \qquad (1.11)$$

where ϵ_0 is the permittivity of free space, equal to $(1/36 \pi) \times 10^{-9}$ F/m. It is seen that k is analogous to the relative permeability μ_r.

It can be shown that the speed c of electromagnetic waves (e.g. light) in a dielectric is given by the expression

$$c = 1/(\epsilon\mu)^{\frac{1}{2}} = 1/(k\epsilon_0\mu)^{\frac{1}{2}} \qquad (1.12)$$

1.4. THE PROPERTIES OF DIELECTRIC MATERIALS

For free space (i.e. a vacuum) c is equal to 3×10^8 m/s. For practical purposes it can be assumed that μ for dielectrics is equal to μ_0. Non-dielectric materials are opaque to electromagnetic radiation.

Equation 1.11 assumes that the space between the plates of the capacitor consists of a single homogeneous medium. However, when a slab of dielectric material of thickness t and permittivity ϵ is inserted between the plates separated by a thickness d of air, the equation for the capacitance becomes, assuming that air has the same permittivity as free space,

$$C = \frac{A}{(d-t)/\epsilon_0 + t/\epsilon} = \frac{\epsilon_0 A}{d - t(1 - 1/k)} \qquad (1.13)$$

It is seen that the right-hand side of equation 1.13 contains two variables, t and k, for a fixed value of d. If one of these is kept constant or has a known value, the value of the other can be determined by measuring C with a suitable capacitance bridge. Details of the different types of capacitance bridge are given by Hague[5].

One type of bridge which is often used for measuring capacitances is illustrated in *Figure 1.11*. Like the inductance bridge, described

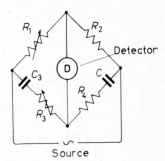

Figure 1.11. A.C. bridge for measuring capacitance

in Section 1.3, the source is an oscillator operating at a frequency of the order of 1,000 Hz and the detector may be a cathode-ray oscilloscope, a valve-voltmeter, a thermo-galvanometer, or a pair of headphones. C is the capacitance under test, C_3 a standard capacitor of suitable value, and R_1, R_2, R_3 and R_4, non-inductively wound resistors of known values, R_1 and R_3 being variable. R_1 and R_3 are adjusted, in turn, until the detector signal is a minimum, for which

$$C = C_3 R_1 / R_2 \qquad (1.14)$$

17

An important advantage of this particular method is that only resistors, which are more reliable in their operation and less costly than variable capacitors, need be varied. In common with all other types of a.c. bridges, care must be taken to ensure that stray capacitances are minimized and that proper earthing is maintained.

If a slab of material of constant thickness is placed between the plates of a capacitor, any deviations of the physical properties of the slab from those of a standard specimen of similar size and shape can be detected from variations in k as obtained by measuring changes in capacitance, using equation 1.13. Alternatively, if it is known that the physical properties of different samples of a given material are constant, variations in thickness from that of a standard sample can be determined by this method. Thicknesses of coatings of paint, rubber, or plastic on a metal surface have been measured in this way by determining the capacitance between the metal surface and a parallel plate placed at a fixed distance from it.

1.4.3. MICROWAVE TECHNIQUES

One of the most recent developments in the field of non-destructive testing of dielectrics is the use of electromagnetic microwaves. This technique has certain advantages in many cases over ultrasonic testing methods and much progress is now being made in its application to both quality control and thickness gauging. It has proved to be particularly useful in the testing of plastics, foodstuffs, rubber, glass, concrete, wood, ceramics, and semiconductor materials.

As explained in Section 5.3.2 the electromagnetic radiation spectrum extends over a wide range of frequencies which embraces radio waves at the lower end and visible light, x-rays, and γ-rays towards the upper end. In a vacuum these waves travel a ta speed of 3×10^8 metres per second but, in materials, they travel at lower speeds, depending on the values of their magnetic permeabilities μ (usually equal to the value μ_o for free space, i.e. $4\pi \times 10^{-7}$ H/m, for a dielectric) and electrical permittivity ϵ, and thus the dielectric constant k, as given by equation 1.12.

With the established techniques used for ordinary radio and television broadcasting, low frequency electromagnetic waves are generated by the electrical oscillations in a tuned circuit. Here the frequency range extends from about 150 kHz to just over 600 MHz and the wavelengths in free space corresponding to these extremities of frequency are 2,000 m and 50 cm, respectively. However, in recent years, the advancement of radar techniques has led to the use of devices for radiating waves with frequencies extending from 10,000 MHz (i.e. 10 GHz) to 100,000 MHz (i.e. 100 GHz) for which the corresponding wavelengths in free space are 3 cm and 3 mm, respectively. The term *microwaves* is used to describe this kind of radiation.

Microwaves are produced by high frequency electronic oscillators of which there are a number of different types (see, for example, Sims and Stephenson[10]). The source most suitable for testing purposes is the klystron tube, which is reliable, relatively cheap, and easily obtainable. Unlike low frequency alternating currents, currents generated at microwave frequencies cannot flow along wires. Propagation is possible only in specially constructed waveguides in which the waves are channelled without significant loss and conveyed from the oscillator to the probe. The probe usually consists of a sheet-metal funnel or horn placed in a suitable position in relation to the test sample (see *Figure 1.12*). A similar probe receives the waves after they have passed through the material. For detection, a small but highly sensitive device, such as a crystal diode or thermistor, is used. The output is amplified and measured with, say, a valve-voltmeter or cathode ray oscilloscope. Provision is made in the waveguides for inserting calibrated attenuators and phase shifting devices.

The propagation of microwaves in a medium is governed by the dielectric constant k and the dissipation factor (or dielectric loss). The dielectric constant not only determines the wave velocity (see equation 1.12) but also plays the same part in electromagnetic propagation as the characteristic impedance plays in ultrasonics. Reflections occur at discontinuities in dielectric constant, such as those produced by boundaries between two media, and by defects. The dissipation factor corresponds to the absorption coefficient in ultrasonics.

Microwave techniques can be used for the measurement of distances, thickness gauging, investigating the physical properties of materials, and flaw detection. They have certain advantages over ultrasonic methods. First, coupling between different media, whether solid, liquid, or gaseous, is good and the use of special couplants is unnecessary. Second, attenuation for a given distance is very much less than for ultrasonics. Thus materials such as rubbers and plastics, which are high absorbers of ultrasound, are particularly suitable for microwave methods of testing.

Disadvantages of microwave methods are that tests can be conducted only with non-metals and that the use of short pulses, as employed for ultrasonic flaw detection, is not possible. With short pulses, small defects in solid materials could easily be located. In ultrasonic testing, the propagation of pulse lengths of as short

19

as 2 mm is often possible; a 2 mm pulse of ultrasonic waves in a solid has a duration of from 0·5 to 1μs, depending on the nature of the material. In microwave testing, where the wave velocities are about 10^5 greater than acoustic velocities, a 1 μs pulse has a length of the order of 100 m. Using the latest type of electronic equipment, it is possible to reach a lower limit for pulse duration of about 20 ns $(2 \times 10^{-8} \text{ s})$ for which the pulse length is of the order of 3 m; this is clearly useless for the detection of individual small flaws. The frequency bandwidth for this pulse duration would be 500 MHz.

There are several different methods of testing with microwaves. In one of these (see *Figure 1.12a*) the sample is placed between

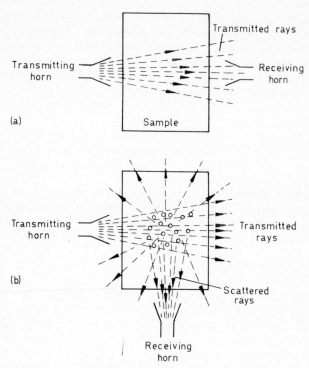

Figure 1.12. Relative positions of probes and sample for microwave testing. (*a*) Transmission measurements. (*b*) Scattering measurements for the detection of porosity

two probes situated opposite one another and the observed change in phase enables one to calculate the thickness of the dielectric (if the wave velocity is known) or the wave velocity (if the thickness is known). The attenuation is determined from the observed fall in level of the detected signal when the sample is placed between the probes. Velocity and attenuation measurements can provide information about the physical properties of the sample, such as purity, moisture content, and structural variations. The presence of extensive defects may be detected from attenuation measurements.

A single-probe technique may be used when the sample is accessible from one side only. For example, the thickness of a non-metallic coating on a metal surface can be measured by means of a reflection interference method. The detector is incorporated in the waveguide system between the horn and source and it probes the standing-wave pattern formed by the waves incident from the source and those reflected from the top and bottom surfaces of the coating. Variations in coating thickness will modify the standing-wave system and induce variations in the intensity of the signal picked up by the receiver. The instrument can be calibrated for known thicknesses by a micrometer phase changer inserted in the waveguide system. Thicknesses can be measured by this method to an accuracy of better than 0·001 inch (25 μm).

Except where defects are sufficiently large to give rise to specular reflections (i.e. several wavelengths across, about 20 mm for 4 mm waves) detection of individual flaws is impossible, for reasons stated above. However, with widespread defects, such as distributed porosity, detection can be made by observing scattered radiation, using the technique illustrated in *Figure 1.12b*.

Fuller accounts of microwave methods of testing are given by Hochschild[11] and Dean[12].

1.4.4. ELECTRIFIED PARTICLE TESTING

In Chapters 3 and 4 magnetic particle methods of testing in ferromagnetic materials are described. Similar types of tests with electrified particles have been made on dielectric materials, one example being the detection of cracks on glazed surfaces such as those of ceramic materials.

The electrified particle method of testing was devised by de Forest and Staats in 1945, originally for the detection of minute cracks in glass bottles. Further developments have since been made by the Magnaflux Corporation, who market equipment for this technique under the trade name *Statiflux*.

There are two main applications of this method, one of which is to the testing of surfaces of dielectric layers on metal backings and the other is to the testing of the surfaces of dielectric materials unbacked by any conducting substance. In both cases, positively charged minute particles of calcium carbonate are applied from an aerosol spray to the surface under test. When a dielectric layer is backed by a metal, free negative electric charges in the metal are attracted by these positive charges and appear at the boundary surface (see *Figure 1.13*) so that local electric fields are induced.

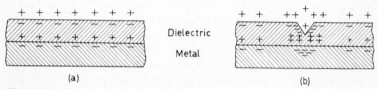

Figure 1.13. Charge distribution on dielectric and metal surfaces with electrified particle testing. (*a*) Defect-free sample. (*b*) Sample with surface crack on dielectric layer

Where a crack occurs on the surface, the thickness of the dielectric layer at the defect is reduced and the intensity of the local electric field is increased. The free negative charges on the metal surface tend to accumulate at the base of the crack, thus attracting to the upper surface of the dielectric an increased density of the positively charged particles in such a way that a magnified outline of the flaw is observed.

When the material under test is not backed by a conductor, free charges may be introduced in the following manner. Many wetting liquids, including tap water, are electrically conducting because they contain free charges, i.e. ions. If the surface under test is wetted by such a liquid and then dried, some of the liquid will be retained in any cracks which may be present. On spraying the surface with the positively charged calcium carbonate particles, the free charges in the liquid deposited in the defects will have the same effect on the particles as those on a metal surface backing a dielectric layer. Thus, as before, the particle density will be greatest in the vicinity of a crack, the magnified outline of which will be observed.

The electrified particle method can be highly sensitive, especially for the detection of fine cracks, and magnifications of the order of 30,000 × are possible. This compares favourably with the penetrant technique described in Chapter 6. Applications include the detection of cracks in glass lenses, glass-metal seals, and false teeth as well as those in glass containers and ceramics.

22

REFERENCES

Further information on electrified particle testing is given by McMaster[13].

REFERENCES

1. SCOTT, W.T. *The Physics of Electricity and Magnetism*, Wiley, New York, 1959.
2. VALDES, B. *Proc. I.R.E.*, New York, 1954, **42,** 420.
3. PERRY, C. C. and LISSNER, H. R. *Strain Gage Primer*, 2nd Edn., McGraw-Hill, New York, 1962.
4. DE MEESTER, P., DEKNOCK, R., and VERSTAPPEN, G. *Mater. Eval.* 1966, **24,** 482.
5. HAGUE, B. *Alternating Current Bridge Circuits*, Pitman, London, 1930.
6. MIX, P. E. *Mater. Eval.* 1966, **24,** 253.
7. McMASTER, R. C. (Ed.). *Nondestructive Testing Handbook*, Ronald, New York, 1959, Vol. II, Section 34.20.
8. McMASTER, R. C. (Ed.). *Nondestructive Testing Handbook*, Ronald, New York, 1959, Vol. II, Section 42.23.
9. McCAIG, M. *Progress in Non Destructive Testing* (Eds. Stanford, E. G. and Fearon, J. H.), Heywood, London, 1961, Vol. 3, p. 131.
10. SIMS, G. D. and STEPHENSON, I. M. *Microwave Tubes and Semiconductor Devices*, Blackie, London, 1963.
11. HOCHSCHILD, R. *Nondestruct. Test.* 1963, **21,** 115 and *Mater. Eval.* 1968, **26,** 35A.
12. DEAN, D. S. *Non Destruct. Test. (U.K.)* 1967, **1,** 19.
13. McMASTER, R. C. (Ed.). *Nondestructive Testing Handbook*, Ronald, New York, 1959, Vol. II, Sections 28 and 29.
14. KOVACS, I., NAGY, E., and FELTHAM, P. *Phil. Mag.*, 1964, **9,** 797.

23

2

EDDY CURRENT METHODS

2.1. INTRODUCTION

THE use of eddy currents for non-destructive testing was pioneered by F. Förster and his associates[1] in Germany and further developments were made in Great Britain by E. G. Stanford[2] and in the U.S.A. by R. Hochschild and others[3-7]. Eddy current tests can be carried out on all materials which conduct electricity. They have been used for crack detection, for thickness measurements of metallic plates, foils, tubes and cylinders and of non-metallic coatings on metals, for the detection of corrosion, for measurements of conductivity, for the investigation of chemical compositions of metals, and so on. A great advantage of the method is that direct contact with the test-piece is not necessary.

Eddy currents may be induced in an electrically conducting body by placing the latter near a coil through which an alternating current is flowing, in the same way as a current is induced in the secondary coil of a transformer when the primary is excited. Indeed, the induction of eddy currents in transformer cores often proves troublesome owing to the generation of unwanted heat.

For the application of eddy currents to non-destructive testing, an alternating current of a given frequency is generated in the *primary* or *exciting* coil A (see *Figures 2.1a* and *b*). An alternating magnetic flux N_1 is produced and this induces an alternating current of the same frequency in the coil B, which may be described as the *secondary*, *pick-up*, or *search* coil. With the introduction of the specimen C, the alternating flux N_1 induces in the latter an eddy current flow which gives rise to an alternating magnetic flux N_2 in the opposite direction; the current in the secondary coil B is consequently reduced. For given conditions, the reduction in current should be equal for all identical specimens placed in the same position relative to the coils. Any observed inequality in the value of the reduced current could indicate the presence of a defect, a change in dimensions, or a variation in the electrical conductivity or in the magnetic permeability of the sample due, perhaps, to a change in its physical or chemical structure.

24

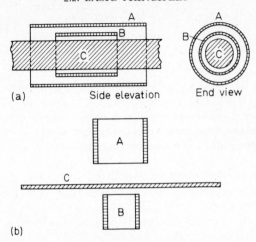

Figure 2.1. Arrangements of exciting coil (primary) A, search coil (secondary) B, and test sample C for eddy current testing (*a*) cylindrical specimen, (*b*) sheet specimen

2.2. BASIC PRINCIPLES

2.2. GENERAL CONSIDERATIONS

For the sake of simplicity we shall consider the arrangement depicted in *Figure 2.1a* where the test-piece is cylindrical and lies coaxially with the primary and secondary coils. The treatment can be extended to apply to other configurations.

A current in the primary coil will induce a magnetic field inside it. Let H_0 represent the r.m.s. value of the intensity of the magnetic field induced within the secondary coil in the absence of the specimen, i.e. when the coil contains only air. Provided that the coil is long enough, H_0 will be constant over a given cross-section sufficiently far removed from its ends. The corresponding r.m.s. flux density B_0 is given by

$$B_0 = \mu_0 H_0 \qquad (2.1)$$

where μ_0 ($= 4\pi \times 10^{-7}$ H/m) is the permeability for free space and, effectively, for air and almost all other non-ferromagnetic materials. When the electrically conducting test-piece is inserted within the secondary coil, there is an observed change in the r.m.s. flux density, which is no longer constant for all points in a given cross-section because of the non-uniformity of the eddy-current distribution (see *Figure 2.6*). We shall consider a *mean* r.m.s. flux density $\overline{B}$ over the cross-section, defined as follows

$$\overline{B} = \mu_r \mu_o \overline{H} \tag{2.2}$$

where $\overline{H}$ is the mean r.m.s. field strength and μ_r the relative permeability of the specimen (see Section 1.3.1.). Separate consideration will be given to the cases where μ_r is approximately equal to unity, i.e. for non-ferromagnetic conductors and where μ_r has a value greater than unity, i.e. for ferromagnetic materials.

2.2.2. Non-ferromagnetic Conductors

Let us assume that the r.m.s. field strength H_0 remains unchanged on the introduction of the specimen inside the secondary coil and, instead, that the relative permeability has changed from unity to a value $\overline{\mu}$ given by

$$\overline{B} = \overline{\mu}\mu_o H_o = \overline{\mu} B_o \tag{2.3}$$

where $\overline{\mu}$ is called the *effective permeability*. This quantity has two components*, one being $\overline{\mu} \cos \phi$ and the other $\overline{\mu} \sin \phi$, where ϕ is the *phase angle*. The two components are said to be 90° out of phase with one another. Because the distribution of eddy currents in a specimen is a function of frequency (see *Figure 2.6*), $\overline{\mu}$ must vary with frequency. *Figure 2.2* illustrates the variation of the two components of $\overline{\mu}$ with frequency as calculated by Förster[1] from theoretical considerations. It is assumed that the specimen completely fills the secondary coil (i.e. the fill factor or area efficiency is 100 per cent). The numbered points marked on the curve represent values of the parameter f/f_g, where f is the frequency of the exciting current and f_g is the *limiting* or *boundary frequency*, as defined by the following expression

$$f_g = 2/(\pi \mu_o \sigma D^2) = 5{,}066/(\sigma D^2) \tag{2.4}$$

where σ is the electrical conductivity in m/ohm. mm² and D the diameter in cm for the specimen. Values of f_g for cylindrical rods 1 in (25 mm) diameter for a number of materials are given in *Table 2.1*. Because the quantity f_g is characteristic of both the conductivity and the diameter of the specimen, the graph shown in

---

* In order to remain on an elementary plane, the argument has been simplified. Strictly speaking, $\overline{\mu}$ is a complex quantity given by

$$\overline{\mu} = |\overline{\mu}| \cos \phi + j\,|\overline{\mu}| \sin \phi = |\overline{\mu}| \exp (j\phi) \tag{2.3a}$$

where $j = (-1)^{\frac{1}{2}}$ and $|\overline{\mu}|$ is the magnitude of μ.

Figure 2.2 will be identical for all materials for which the relative permeability is equal to unity.

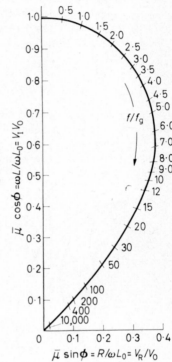

Figure 2.2. Curve showing the variation of the components of effective permeability (normalized impedance and voltage) with frequency (Förster[1]) (*By courtesy of Forster Instruments Ltd.*)

Table 2.1. Approximate values of electrical conductivity σ, limiting frequency f_g and optimum frequency $f = 15f_g$ for metal rods 1 in (25 mm) diameter at room temperature

Material	Conductivity σ m/ohm mm^2	f_g Hz	$f = 15f_g$ kHz
Aluminium	40	20	0·30
Copper	60	13	0·20
Iron (saturated)	12	65	0·97
Nickel	16	50	0·75
Tin	9	87	1·3
Zinc	20	40	0·60

27

In practice one does not actually measure $\bar{\mu} \cos \phi$ and $\bar{\mu} \sin \phi$. One measures either the impedance of the coil or the voltage across it. The impedance has two components, one being resistive, given by R, the measured resistance of the coil, and the other reactive, given by ωL ($\omega = 2\pi f$), where L is the measured inductance of the coil. Similarly, the voltage V has two components V_R and V_I, respectively. It can easily be seen that:

and
$$\bar{\mu} \cos \phi = \omega L / \omega L_o = V_I / V_o \qquad (2.5a)$$
$$\bar{\mu} \sin \phi = R / \omega L_o = V_R / V_o \qquad (2.5b)$$

where L_o is the inductance of the coil and V_o the voltage across it when the specimen is absent. The ratios $\omega L / \omega L_o$ and $R / \omega L_o$ are called the *normalized* impedance components and V_I / V_o and V_R / V_o the normalized voltage components; these are independent of the properties of any particular coil.

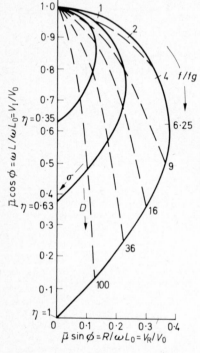

Figure 2.3. Effective permeability (normalized impedance and voltage) curves for different fill factors η. Dashed lines join points representing equal values of the parameter f/f_g (Förster[1]) (*By courtesy of Forster Instruments Ltd.*)

Because the cross-section of the specimen can never completely fill that of the secondary coil, the term *area efficiency* or *fill factor* must be introduced. This is defined as the ratio of the cross-sectional area of the specimen to the effective cross-sectional area of the secondary coil, i.e.

$$\eta = (D/D')^2 \tag{2.6}$$

where D is the diameter of the specimen and D' is the inside diameter of the coil. *Figure 2.3* shows variations of impedance with frequency for different values of η. The dashed lines join all points on the curves corresponding to similar values of f/f_g. Thus continuous lines represent variations of conductivity σ for constant diameter D and the dashed lines represent changes in D for constant values of σ. Using suitable phase-sensitive apparatus, one can measure changes of impedance or voltage due to these variations. The relative magnitudes of the impedance and voltage changes with D and σ depend on the frequency. *Figure 2.4* illustrates curves obtained

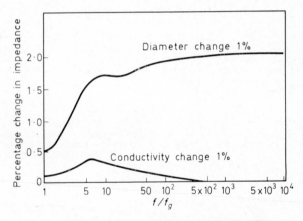

Figure 2.4. Percentage changes in impedance for (*i*) a diameter change of 1 per cent with constant conductivity and (*ii*) a conductivity change of 1 per cent with constant diameter (Förster[1]) (*By courtesy of Forster Instruments Ltd.*)

by Förster for one per cent variations in D and σ, respectively. It is seen that diameter changes are more perceptible than variations in conductivity and also that changes in conductivity produce greater variations of impedance at lower frequencies than at higher frequencies.

29

2.2.3. FERROMAGNETIC CONDUCTORS

For ferromagnetic materials, such as mild steel, cast iron, and nickel, account must be taken of the relative permeability μ_r which will be greater than unity. The limiting frequency f_g must be defined, more exactly, as follows

$$f_g = 2/(\pi\mu_r\mu_0\sigma D^2) = 5,066/(\mu_r\sigma D^2) \qquad (2.4a)$$

When the specimen is kept magnetized to well above saturation level by the application of a sufficiently high direct magnetic field, the relative permeability is approximately unity and remains constant at that value throughout the test for not too large changes in the magnetic flux. The curves depicted in *Figures 2.2, 2.3* and *2.4* are thus applicable.

When the specimen is not saturated magnetically, the Förster curves can still be plotted in a satisfactory manner, provided that

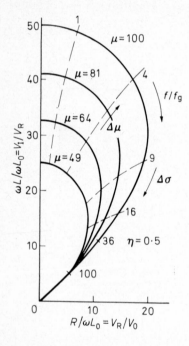

Figure 2.5. Normalized impedance (and voltage) curves for ferromagnetic cylinders (Förster[1]) (*By courtesy of Forster instruments Ltd.*)

the eddy currents are sufficiently low to produce only small periodic relative changes in μ_r. The impedance of the secondary coil will then be affected by three variables, namely μ_r, σ and D. Provided that the value of μ_r is sufficiently high (e.g. greater than 50), the

30

influences of the fill factor η and hence the diameter D of the specimen are negligible and the effective variables are simply μ_r and σ. *Figure 2.5* illustrates a set of curves obtained by Förster for $\eta = 0.5$. The continuous lines represent variations in σ for fixed values of μ_r and the dashed lines indicate changes in μ_r for given values of σ. Thus the eddy current method is not sensitive to small variations of dimensions for ferromagnetic materials, unless they are magnetized to saturation.

2.2.4. PENETRATION OF EDDY CURRENTS

It is found that the eddy current intensity is greatest at the surface and decreases as the depth in the specimen increases. *Figure 2.6*

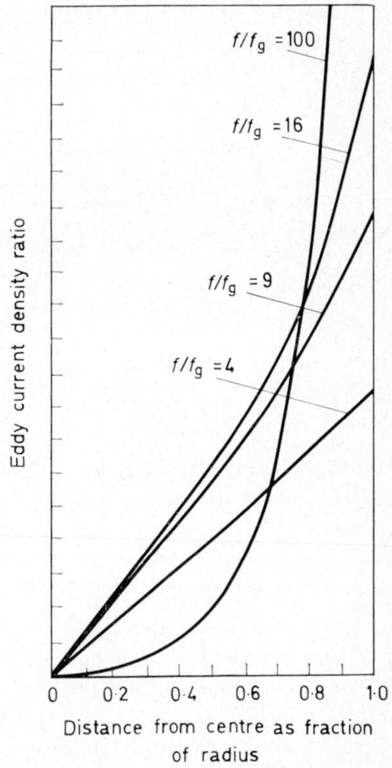

Figure 2.6. Eddy current distribution in a cylindrical bar for different frequencies (Hochschild[4]) (*By courtesy of the Society for Nondestructive Testing Inc.*)

illustrates how the ratio of the eddy current density at a given point to that at the surface varies with the radial distance from the axis of a cylindrical specimen. This is of great significance for the choice of frequency used in flaw detection, because it is not possible to detect internal cracks at very high frequencies. The tendency of the currents to remain in the neighbourhood of the surface at high frequencies is known as the *skin effect*, a phenomenon well known to radio-frequency and microwave engineers.

For practical purposes, the term *penetration depth* is used to determine the range of eddy current activity within a material at a given frequency. For cylinders, this is given by the ratio $(r - r')/r$, which is expressed as a percentage. r is the radius of the cylinder and r' the radius of the zone in which the eddy current density is less than $1/2 \cdot 718$ (i.e. $1/e$) of its value at the surface. The penetration depth will thus depend on the ratio f/f_g. Values at different frequencies may be obtained by using a cylinder in which is bored a conical cavity (see *Figure 2.7*). A search coil, placed to fit closely around the

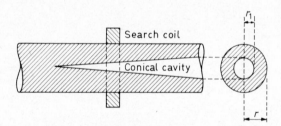

Figure 2.7. Conical cavity arrangement used by Förster[1] for determining penetration depths (*By courtesy of Forster Instruments Ltd.*)

exterior, is moved along the cylinder until the impedance decreases to $1/2 \cdot 718$ of its original value. The radius of the cavity at this point is r'.

2.2.5. THE EFFECTS OF FLAWS ON EDDY CURRENTS

Any defect, such as a crack, will affect simultaneously the values of both σ and D for the material. Thus variations of impedance due to the presence of defects will, in general, lie in directions other than along the constant σ or constant D curves. Förster[1] has investigated the effects of cracks and blow-holes in cylindrical specimens on the impedance of the pick-up coil. He used, as a model, a tube filled with mercury into which he introduced pieces of plastic to simulate various types of cracks and blow-holes. These defects were classified as to location, size, and origin and identified by symbols (see *Figure 2.8*). Given values of f/f_g were selected and the effects of different kinds of flaw were studied. He found that useful results could be obtained only where f/f_g lies between 5 and 150. Where f/f_g is greater than 150, the sensitivity of crack detection falls off rapidly for comparatively small changes of diameter and where

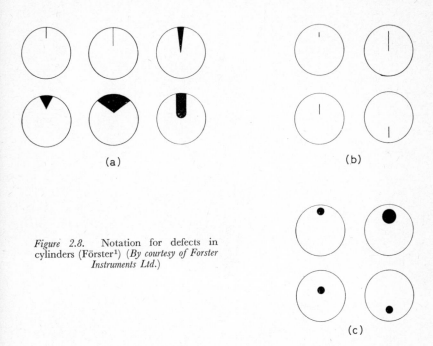

(a)

(b)

(c)

Figure 2.8. Notation for defects in cylinders (Förster[1]) (*By courtesy of Forster Instruments Ltd.*)

f/f_g is less than 5, the angle between the directions of the ΔD curve and the curve representing crack variations is too small for the two components to be resolved with any accuracy. The optimum value of f/f_g is in the region of 15 where the component of impedance perpendicular to the ΔD curve reaches its maximum value for a given crack.

Figure 2.9 shows a section of the impedance graph for $\eta = 1$ in the neighbourhood of the point $f/f_g = 15$, which is taken as the origin. The two components $\Delta (\omega L)/\omega L_0$ and $\Delta R/\omega L_0$ are plotted. The lines OA and OB represent ΔD and $\Delta \sigma$ respectively. The circle centred at the origin shows that a 1 per cent change in D produces the same variation in the magnitude of the impedance as a 7 per cent change in σ. The anticlinal curve (i.e. convex upward) OC represents changes in impedence for a thin surface crack of a depth-to-width ratio of 100 or more in the radial direction, the numbers indicating the depth of the crack as a percentage of the diameter.

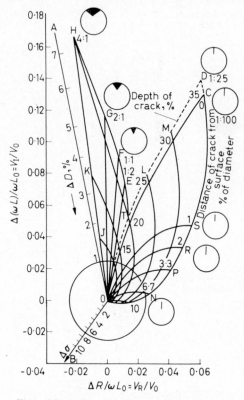

Figure 2.9. Effects of changes of diameter and conductivity, and the presence of cracks on the normalized impedance (and voltage) for $f/f_g = 15$ (Förster[1]) (*By courtesy of Forster Instruments Ltd.*)

The dashed curve OD represents a similar crack of depth-to-width ratio 25. The almost straight lines OE, OF, OG, and OH represent radial cracks which have width-to-depth ratios 1:2, 1:1, 2:1, and 4:1. The synclinal curves (i.e. concave upward) OJ, OK, OH, OL, OM, and OC represent cracks, not necessarily starting from the surface, which have constant length-to-diameter ratios, and the anticlinal curves ON, OP, OR, and OS denote the depth of the top of the crack from the surface as a percentage of the diameter.

34

Figure 2.10 shows the two curves OC of *Figure 2.9*, for various values of f/f_g, superimposed on the complete impedance curves. The short curves marked with the ratios up to 4:1 correspond to the curve TEFGH in *Figure 2.9* and the dashed curve represents a

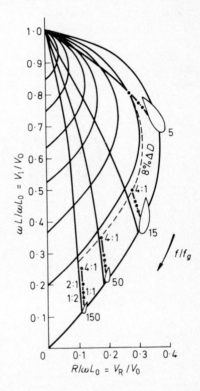

Figure 2.10. Curves of *Figure 2.9* in relation to the complete impedance diagram (Förster[1]) (*By courtesy of Forster Instruments Ltd.*)

decrease of 8 per cent in diameter. Förster has also obtained curves corresponding to tubular specimens for which he took into account variations in wall thickness and departures from cross-sectional symmetry (see, for example, Hochschild[3] and McClurg[7].)

2.3. METHODS OF MEASUREMENT

2.3.1. COIL ARRANGEMENTS

A number of different coil arrangements are possible with eddy current testing, each dependent on the type of application and the required degree of sensitivity. The highest sensitivity is attainable with the use of two coils, a large primary coil for excitation and a smaller secondary coil for detection. However, the design is simplified if the same coil is used for both excitation and detection. With this latter arrangement there is a decrease in output but it may be no disadvantage with the use of highly sensitive modern electronic equipment. It is sometimes found, however, that the employment of separate exciting and pick-up coils provides better impedance matching.

Three different types of coil are generally employed, these being the *encircling*, the *internal probe*, and the *external probe* types (see *Figure 2.11*). Encircling coils are used for the testing of small diameter

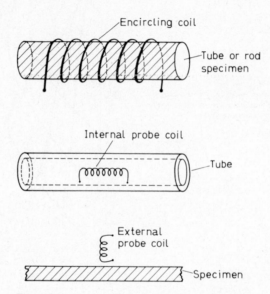

Figure 2.11. Various coil positions for eddy current testing (the same coil is used for both generation and detection)

rods, tubes, and spheres. Internal coils are applied to the inspection of the inner surfaces of small diameter tubes; the windings of these coils are coaxial with the tubes being examined. External probe coils are used for the testing of objects with plane or small curvature surfaces, e.g. large diameter tubes and rods; they are positioned with their axes perpendicular to the surfaces of the test specimens. An advantage of the use of probe coils is that they can be made with very small diameters (down to 1 mm) and thus detect the presence of very small flaws. The sensitivity of the coil can be increased by the insertion of high permeability cores such as ferrite rods. These small sensitive coils are often called *focused coils*.

Förster[1] has described three main coil arrangements (see *Figure 2.12*). It is seen that in the absence of any test specimen, the secondary voltages across AB and BC, respectively are equal and opposite, provided that the coils are properly matched; no resultant voltage is then detected across AC. The introduction of the test-piece (*Figure 2.12a*) results in a change in impedance between B and C. The opposing secondary voltages are no longer equal and a measurable voltage appears across AC. The use of this arrangement is called the *absolute method*.

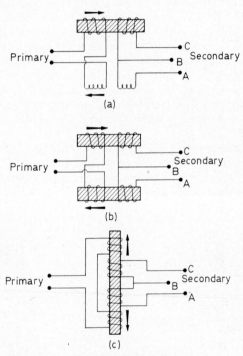

Figure 2.12. The three principal coil arrangements used by Förster[1] (*By courtesy of Forster Instruments Ltd.*)

One is generally interested in variations in the physical properties of the specimen, e.g. the existence of flaws and changes in conductivity due to impurities, etc. These variations normally produce only small changes of impedance and voltage compared with the large changes caused by the introduction of the specimen. With the absolute method the sensitivity may be too low. A large increase in sensitivity can be obtained with the *comparison method (Figure 2.12b)* which consists of the use of two identical coil assemblies. A standard defect-free specimen is placed in one coil and the test sample in the other; changes arising from the differences between the two samples are measured. The method suffers the disadvantage in that comparatively small changes in structure or chemical composition may give rise to greater variations in impedance than changes due to the presence of defects. This difficulty can be overcome by the use of the *auto-comparison method* illustrated in *Figure 2.12c.* Here two different parts of the same sample are compared with one another. The method is commonly used for the testing of wires and long tubes and bars which are fed through the coils at a constant speed. Indications due to the presence of defects change more rapidly than those caused by dimensional and conductivity variations. If the output is differentiated by the use of a suitable circuit, the indications are greatly increased in magnitude. When a longitudinal crack passes through both pick-up coils at the same time, no indication is observed, but the ends of this type of defect can easily be detected.

Similar arrangements can be made when a single coil is used for both excitation and detection and also when probe type coils are employed. *Figure 2.13* illustrates two alternative arrangements for the auto-comparison method using two single-probe coils.

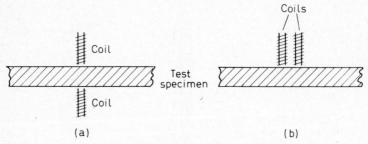

Figure 2.13. Alternative positions of external probe coils with the autocomparison method

2.3.2. The Suppression of Undesirable Effects

For conductivity investigations, effects due to dimensional changes in the test sample must be eliminated and, conversely, for dimensional measurements, conductivity effects must be suppressed.

If, at a given frequency, conductivity changes only are to be determined, effects due to variations in dimensions can be eliminated by measuring the component of impedance (or voltage) in the direction perpendicular to the constant D curve in *Figure 2.3*, using a suitable phase sensitive detector. Förster (see McMaster[8]) incorporated a phase controlled rectifier in the designs of the Multitest and Sigmaflux instruments. A control voltage with a phase 90 degrees to that corresponding to the undesired effect is generated. An arrangement of rectifiers in a bridge circuit ensures that only that component of the voltage across the test coil which is in phase with the control voltage is measured. For dimensional investigations, the effects due to conductivity can be greatly reduced by operating at a high frequency (see *Figure 2.4*).

When a probe type coil is used for detection, there is a further undesired effect due to 'lift-off' (McMaster[8]). When the probe coil is lifted off the surface of the test sample, a change occurs in the impedance of and, hence, the voltage across the coil. At a given frequency, this variation may be represented by a line along a direction other than that of either the ΔD or $\Delta \sigma$ curves of the Förster diagram. The magnitude of the displacement along this curve depends on the height by which the coil has been raised from the surface of the specimen. Lift-off can be compensated by the application of a voltage of equal and opposite magnitude and phase.

The lift-off effect occurs when a sample having its surface covered with a non-metallic substance, such as paint, is tested with a probe coil. The effect can be compensated in this case by observing the change of indication when a probe is placed on the surfaces of two similar defect-free samples, one coated and the other uncoated. Compensation for lift-off is incorporated in the design of Förster's Defectometer. With the Ultrasonoscope Surface Film Thickness Meter, the lift-off effect is actually utilized for measuring thicknesses of surface films and coatings.

2.4. EDDY CURRENT TESTING EQUIPMENT

2.4.1. INTRODUCTION

A wide variety of eddy current testing equipment exists but the limited space available here restricts the choice for description to only a few items, which are intended to be representative. The criterion for selection is the familiarity of the apparatus to the authors, who wish to acknowledge those manufacturers of first-class equipment whose names have not been mentioned.

39

2.4.2. FÖRSTER'S MULTITEST APPARATUS

Förster's analysis has been applied directly to the design of his Multitest equipment which employs the comparison method described in Section 2.3.1. and illustrated in *Figure 2.12b*. It is a versatile instrument which can be used for conductivity testing, the investigation of dimensional variations, and flaw detection. The two components V_R and V_I of the voltage across AC are separated in phase and fed to the X- and Y-plates, respectively, of an oscilloscope, on the screen of which appears a bright spot representing a point on the Förster impedance analysis graph (*Figures 2.3, 2.9,* and *2.10*). With suitable amplifiers and attenuators the scale of the display can be varied as required. For flaw detection the scale is magnified so that one can, for example, simulate the curves shown in *Figure 2.9*. For a defect-free rod of given dimensions and conductivity, a point O can be selected by operating at a suitable frequency. If the spot moves to the point corresponding to, say, L on the graph when the defect-free rod is replaced by another rod of similar conductivity and dimensions, a defect (in this case a surface crack of depth to diameter ratio of 25 per cent) is indicated as being present.

Conductivity measurements can be made independently of dimensional changes by adjusting the phase control in such a manner that the dimensional variations cause the spot to move in a horizontal direction. Thus, for a defect-free specimen, any vertical displacement of the spot would indicate a conductivity change.

The instrument can be applied to automatic testing. For example, components can be sorted out in three batches in terms of hardness, a property which is related to conductivity, at a rate ranging from 3,000 to 11,000 parts per hour, depending on the nature of the component.

2.4.3. FÖRSTER'S SIGMAFLUX EQUIPMENT

Förster's Sigmaflux, formerly known as the Conductiflux in the United Kingdom, uses the ellipse method[1] for crack detection in tubes and rods. Here again, the comparison method described in Section 2.3.1 and illustrated in *Figure 2.12b* is used for the coil arrangement. A reference voltage, in phase with the signal applied to the primary coil, is fed to the X-plates of a cathode ray oscilloscope and the output voltage across AC is fed to the Y-plates. Now two vibrations at right-angles to one another produce a Lissajous

(Bowditch) figure (see, for example, Blitz[9]). In this instance the vibrations are alternating voltages and, because they are sinusoidal and of the same frequency, the figure is an ellipse. The shape of this ellipse depends on the phase difference between the two voltages and, hence, the phase angle ϕ of the impedance (see equations 2.5a and b). *Figure 2.14* illustrates how the various shapes correspond to different types of crack. The ellipse degenerates to a straight line for a crack-free specimen.

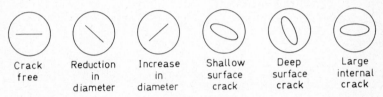

| Crack free | Reduction in diameter | Increase in diameter | Shallow surface crack | Deep surface crack | Large internal crack |

Figure 2.14. Examples of traces observed by the ellipse method (Förster[1]) (*By courtesy of Forster Instruments Ltd.*)

This device can be used for testing ferromagnetic as well as non-ferromagnetic materials provided that a d.c. magnetic saturation unit is used.

2.4.4. C.N.S. CENTEST DIFFERENTIAL COIL INSTRUMENTS

(a) *The Centest 710*

The Centest 710, manufactured by C.N.S. Instruments Ltd., is used for testing tubes, rods, and bars which are passed through an encircling coil assembly at a steady speed of up to 600 ft/min (200 m/min). Higher testing speeds are possible but this entails loss of sensitivity. A simplified block diagram of this instrument is shown in *Figure 2.15*. The test coil assembly consists of two single coils of slightly different impedances placed next to one another and wound in opposite directions. They are excited by an oscillator having stabilities of frequency and amplitude both to within $\pm$ 2 per cent. The impedances of these coils are balanced by two comparison coils (L balance) and a potentiometer device (R balance) located within the main body of the apparatus. The effect of eddy currents on the test coils is to produce two opposing out-of-balance signals, the resultant of which passes through an amplifier, a phase sensitive detector, and a filter to an output stage. A reference voltage supplied by the oscillator to the phase sensitive detector enables one to 'phase-out' unwanted components, e.g. small dimensional variations. The filter rejects interference from both the mains and adjacent machinery.

A device, which can be incorporated in the equipment, uses a photocell arrangement to detect the passage of the end of the tube or rod through the coil and apply a signal to suppress what would otherwise be a spurious indication. For testing ferromagnetic materials, a magnetic saturation unit which reduces the relative permeability to unity (see Section 2.2.3) is used.

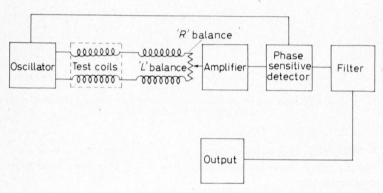

Figure 2.15. Block diagram of the Centest 710 (*By courtesy of C.N.S. Instruments Ltd.*)

There are two alternative operating frequencies, 5 and 10 kHz, chosen to suit the testing of high conductivity materials such as copper. However, a modified version of this instrument, the Centest 710/HF, has been designed for the testing of lower conductivity materials such as stainless steel and other alloys. It can be operated at frequencies of 10 and 50 kHz. *Figure 2.16* illustrates the application of the Centest 710 to tube testing.

The output of this instrument can be linked with a high-speed pen recorder. When a defect in the tube passes through the coils there is an abrupt out-of-balance signal, in contrast to the more gradual changes due to conductivity variations. The variations of the indications on the chart can be easily related to the corresponding portions of the tube under test but the makers have made available a spray indicator which marks the position of the defect on the tube. This unit operates automatically from a triggering device actuated by the sharp signal resulting from the defect.

(*b*) *The C.N.S. Centest 700*

The Centest 700 is a smaller instrument than the Centest 710 and
operates on a slightly different principle; it is suitable for manual

Figure 2.16. Tube testing with the Centest 710 (*By courtesy of C.N.S.
Instruments Ltd. and the Coventry Tube Co. Ltd.*)

testing and can be used with encircling coils, internal probes, and surface type probes. One of the authors (D.G.R.) has applied this instrument to the inspection of 36 inch (914 mm) long and 1 inch (25 mm) o.d. torque tubes of wall thicknesses 10, 12, 16, and 18 s.w.g., at a frequency of 10 kHz (see *Figure 2.17a*). A special saddle-type probe, manipulated by hand, was designed for this purpose. It consisted of two coils, wound in the manner shown in *Figure 2.17b* and connected differentially in the same way as the coils for the Centest 710.

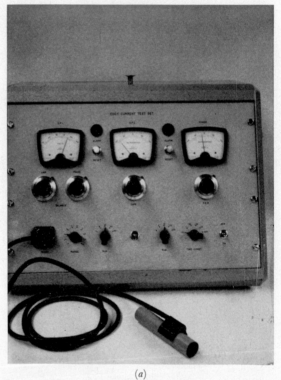

(*a*)

Figure 2.17 (a). The saddle-type probe used with the Centest 700. (*By courtesy of Messrs. Rolls-Royce Ltd., Small Engine Division*)

Because the resultant flux was parallel to the surface one would not have expected a high intensity of induced eddy currents in the material. However, the device proved highly successful in locating surface cracks caused by stress corrosion. Eight passes of the probe along each tube were required for complete scanning. Because of symmetry, unwanted out-of-balance signals which would normally result from the probe being placed on the tube or being removed

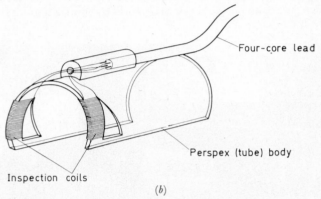

Four-core lead

Perspex (tube) body

Inspection coils

(*b*)

Figure 2.17 (b). Design of the saddle-type probe (*By courtesy of Messrs. Rolls-Royce Ltd., Small Engine Division*)

from it were eliminated, since the changes in coil impedances resulting from these actions were equal and opposite. For the same reason, no out-of-balance signals were observed for a tube coated with an exceptionally thin or thick layer and for tubes of different wall thicknesses.

2.4.5. AUTOMATION INDUSTRIES RADAC EQUIPMENT

The technique of modulation analysis, described by Hochschild[10] is applied to the design of RADAC eddy current equipment. This is manufactured by Automation Industries Inc. (formerly Budd Inc.) and used for testing tubes and rods in continuous motion. The velocity of the component relative to the coil and the changes in the impedance of the coil combine to give rise to a modulation of the operating frequency when variations of eddy current distribution occur. Conductivity changes produce a low frequency modulation and dimensional changes (also permeability variations for unsaturated ferromagnetic materials) cause a somewhat higher frequency modulation. The sharp discontinuities produced by defects such as cracks and blow-holes give rise to modulations at very much higher frequencies.

The sensitivity of the method is increased by the use of 'focused' coils (see Section 2.3.1.), either probe type or encircling. In this way only a small portion of the sample is scanned at any given time.

As with all dynamical measurements, careful control of the speed of testing is essential.

Figure 2.18 illustrates the RADAC 600 used for the automatic testing of aluminium tubing during manufacture. Five different frequencies, namely 500 Hz, 4 kHz, 16 kHz, 64 kHz and 128 kHz,

Figure 2.18. RADAC 600 used for the automatic testing of aluminium tubing during manufacture (*By courtesy of Automation Industries U.K.*)

can be used for operation. The speed of testing may be varied from 10 to 5,000 ft/min (0·25 to 1,250 m/min). The signals are detected with an oscilloscope and a pen recorder. Facilities are provided for the saturation of ferromagnetic components.

Figure 2.19 illustrates recordings on charts for the testing of a given tube at different operating frequencies and using different filters. The tube contained a defect, a periodic variation in dimensions, conductivity variations, and random internal stresses. *Figure 2.19a* shows the response at low frequencies when higher frequency modulations were filtered out; only conductivity changes could be detected. On increasing the frequency (see *Figure 2.19b*), both conductivity and dimensional variations were observed. *Figure 2.19c* shows the indications for higher intermediate frequencies at which only

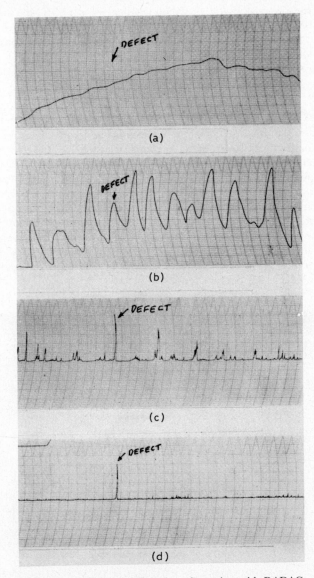

Figure 2.19. Chart recordings for tube testing with RADAC equipment. (*a*) Very low frequencies only. (*b*) Low and very low frequencies. (*c*) Intermediate frequencies. (*d*) Very high frequencies only (*By courtesy of Automation Industries U.K.*)

internal stresses and defects could be identified and *Figure 2.19d* illustrates the response at very high frequencies at which only defects could be observed. For the high frequency measurements relating to *Figures 2.19c* and *d*, the output of the coil was made more sensitive by feeding it to a differentiating amplifier. Thus the effects of conductivity, dimensions and noise were considerably reduced in comparison with those of the defect.

2.4.6. FÖRSTER'S SIGMATEST EQUIPMENT

Förster's Sigmatest, formerly known in the United Kingdom as the Conductitest, uses eddy currents to measure the electrical conductivities of non-ferromagnetic metals to an accuracy of within ± 1 per cent. A single probe coil, acting simultaneously as an exciter and pick-up, is moved by hand over the surface of the test material. The impedance of the coil is initially balanced with that of a similar coil inside the main body of the apparatus. Changes in the impedance of the probe coil due to eddy currents in the material under test give rise to an out-of-balance voltage which is indicated by a meter directly in units of conductivity, i.e. m/ohm. mm^2 and per cent I.A.C.S. (see Section 1.2).

Figure 2.20. Use of the Sigmatest for heat treatment investigations (*By courtesy of Messrs. Rolls-Royce Ltd., Small Engine Division*)

2.4. EDDY CURRENT TESTING EQUIPMENT

The frequency chosen for operation depends on the range of values of conductivity to be measured and the thickness of the material. If the eddy currents do not penetrate the whole thickness of the specimen, the dimensions have no effect on the impedance of the probe coil, except near the edges. Thus an instrument designed for conductivity measurements within the range 5 to 60 m/ohm.mm^2 (10 to 100 per cent I.A.C.S.) is operated at a frequency of 60 kHz for which the depth of penetration ranges from 0·2 mm for $\sigma = 5$ m/ohm.mm^2 to 7·5 mm for $\sigma = 60$ m/ohm.mm^2. Higher frequencies must be used for thinner samples having conductivities within this range. Compensation is provided for 'lift-off' (see Section 2.3.2) when testing materials having rough surfaces.

Applications of the Sigmatest include sorting mixed materials, hardness testing, control of homogeneity, measurement of porosity, and investigating degrees of heat treatment. *Figure 2.20* shows this instrument used for measuring the conductivity of a tensile test-piece to verify that it has been subjected to the correct degree of heat treatment.

2.4.7. THE ULTRASONOSCOPE SURFACE FILM THICKNESS METER

The Ultrasonoscope Surface Film Thickness Meter was originally developed by R. S. Young of the British Aluminium Co. Ltd. It determines the thickness of a non-conducting coating on a non-ferromagnetic metal surface by measuring the 'lift-off' effect for a probe coil (see Section 2.3.2). The probe coil is coupled by a transformer to a tuned circuit which is connected to a highly sensitive and highly stable oscillator of frequency 600 kHz. When the probe is placed in contact with the surface of the coating, the oscillations decrease in amplitude by an amount depending on the coating thickness. The amplitude is then restored to a fixed level, indicated on a meter, by manipulating a potentiometer calibrated in the appropriate units of thickness. The potentiometer readings are zeroed by locating the probe on an uncoated metal surface. Variations of conductivity over a wide range have no effect on the thickness indications.

Two versions of this instrument are manufactured. The first is the ASF/1 which measures anodic film thicknesses on aluminium alloy surfaces within a range from 0 to 0·002 inch (0 to 50 μm) and an accuracy of $\pm 0·25$ μm. The second is the PSF/1 which measures paint thicknesses on non-ferrous metal surfaces within a range from 0 to 0·015 inch (0 to 0·4 mm) and an accuracy of $\pm 0·0001$ inch (0·0025 mm).

The PSF/1 has been used by one of the authors (D.G.R.) to measure paint thicknesses on aircraft castings. Care was taken to avoid both areas of significant curvature and regions near edges, where indications of dimensional effects would have occurred. Variations in readings were observed in places where the painted surfaces were not in a smoothly machined condition and thus mean values of several readings had to be taken. The calibration was checked by means of strips of paper and plastic of known thicknesses which provided simulated paint layers.

2.4.8. Förster's Magnatest Q, QP, and D Instruments[11]

The Magnatest Q was designed by Förster to test ferromagnetic materials by subjecting them to magnetic hysteresis (see Section 1.3.5). Two identical coil assemblies, of either the encircling or probe types, are located at right-angles to one another in order that the flux passing through one set of coils does not pass through the other. The secondary coils are both connected through an amplifier to the Y-plates of an oscilloscope, the X-plates of which are controlled by a time-base (see *Figure 2.21*). An alternating current (e.g.)

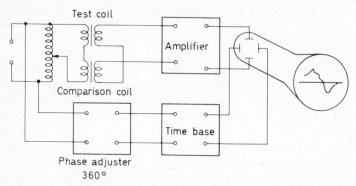

Figure 2.21. Block diagram of the Magnatest Q (*By courtesy of Forster Instruments Ltd.*)

50 Hz mains) is fed through each primary coil in such a way that the two currents are 180 degrees out of phase with one another. The time base can be adjusted so that a single cycle or part of a cycle, of the output from each secondary coil is displayed on the screen. The two signals are superimposed on one another and, in the absence of a test sample, the phases cancel out and a horizontal straight line is observed.

2.4. EDDY CURRENT TESTING EQUIPMENT

When a test specimen is introduced to one of the coils, the material undergoes magnetic hysteresis, the loop of which is modified by the action of induced eddy currents. The straight line becomes disturbed and the trace assumes a shape which is characteristic of the electrical conductivity, the magnetic permeability, and the dimensions of the material. On applying an identical specimen to the second coil in exactly the same relative position, the trace again becomes a horizontal straight line. If, however, the permeability, conductivity, or dimensions of the two specimens differ in any way, the trace assumes a shape which is characteristic of this difference. The Magnatest Q can thus be used to test ferromagnetic components of various shapes and sizes for such properties as hardness, composition, heat treatment, depth of case hardening, the existence of internal stresses, machinability, etc. *Figure 2.22* illustrates samples of

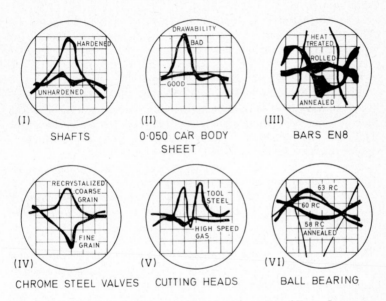

Figure 2.22. Examples of Magnatest Q traces (*By courtesy of Forster Instruments Ltd.*)

traces characteristic of some of these properties. The manufacturers have prepared transparent sheets on which various sample traces were drawn. The sheet containing the trace appropriate to the test is placed in contact with the screen of the oscilloscope and the observed and drawn traces are compared. In the Appendix to this

51

book an account is given of the detailed procedure for the quality control of small components with the Magnatest Q.

The output from the coils of the Magnatest Q may be amplified in the normal manner or passed through either an integrating or differentiating circuit depending on the setting of the controls, i.e. E (normal), $\int E$ (integrating), and D (differentiating). For a small current, the E setting will suffice but, for larger currents which give rise to harmonics and thus produce irregularities in the curve, the $\int E$ setting is required to produce a smoother trace. Greater sensitivity of detection can be obtained with the D setting.

If one is particularly interested in measuring the permeability of the sample, the test can be simplified by the use of the Magnatest QP which is similar to the Magnatest Q but operates at a frequency of only 10 Hz at which eddy current effects are reduced to a minimum.

The Magnatest D is another modification of the Magnatest Q. This utilizes the autocomparison coil arrangement for the testing of bars and tubes for defects.

2.4.9. Bell's Magnetic Reaction Analyser

Magnetic Reaction Analysis, described by McMaster and Smith[12] is an entirely new development in eddy current testing for both non-ferromagnetic and ferromagnetic materials. It makes use of the Hall effect for detection (see Section 1.3.3). The Hall element is highly sensitive to discontinuities and the presence of defects can be accurately determined. Conductivity and dimensional components can be resolved and measured to a high degree of precision.

The Magnetic Reaction Analyser, which applies this technique, is manufactured by F. W. Bell Inc. and has a frequency range 20 to 100 kHz.

The basic principles of the method are illustrated in *Figure 2.23*. The probe consists essentially of an exciting coil and a Hall detector (see *Figure 2.24*). An alternating current of a suitable frequency in the coil induces an alternating magnetic field H_o of constant amplitude in a direction at right-angles to the surface of the test sample. The induced eddy currents give rise to an opposing reaction field H_r. Because of the time delay due to eddy current induction, H_r lags behind H_o in phase by an angle which depends on frequency. The resultant field H_n is the vector sum of H_o and H_r and, in general, is a complex quantity. *Figure 2.25* shows how the magnitudes and phases of H_r and H_n vary with the test frequency. The conductivity and dimensional components can be obtained in the same way as for the Förster curves (*Figure 2.3*).

2.4. EDDY CURRENT TESTING EQUIPMENT

The resultant field H_n is detected by the Hall element, the output of which is fed to the amplifier (*Figure 2.23*). By feeding, at the same time, a subtraction input corresponding to the field H_o to the amplifier, H_r can be determined. A suitable two-channel system permits simultaneous observations of the magnitudes of H_n and H_r.

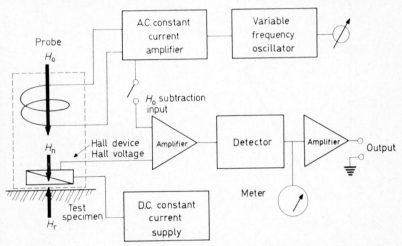

Figure 2.23. Block diagram of the Magnetic Reaction Analyser Model 1090 (McMaster and Smith[12]) (*By courtesy of the Society for Nondestructive Testing Inc.*)

If two Hall probes are placed next to one another on the surface of the test specimen and each connected to a differential analyser, the difference between the fields detected by the probes, can be

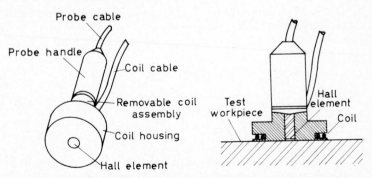

Figure 2.24. Typical example of a surface probe used for the Magnetic Reaction Analyser (McMaster and Smith[12]) (*By courtesy of the Society for Nondestructive Testing Inc.*)

determined. Thus a change of only a small fraction of the reaction field can produce a full-scale deflection and a very high degree of sensitivity of detection is thus obtained.

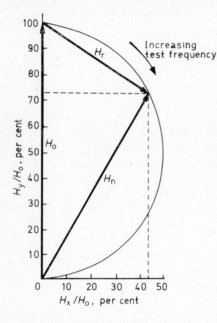

Figure 2.25. Variations of magnetic field vectors with frequency for non-ferromagnetic conductors using magnetic reaction analysis (after McMaster and Smith[12])

The great advantage of magnetic reaction analysis is that, at low frequencies, where the depth of eddy current penetration is a maximum, the magnitude of H_r changes considerably even though H_n varies only slightly with frequency, conductivity, and dimensions. This is in contrast to the correspondingly low sensitivity with more conventional types of eddy current equipment.

REFERENCES

1. FÖRSTER, F. and others. *Theoretische und experimentelle Grundlagen der Zerstörungsfreien Werkstoffprüfung mit Wirbelstromverfahren* and other articles, Z. *Metallk.* 1954, **45,** No. 4; and also McMASTER, R. C. *Nondestructive Testing Handbook*, Vol. II, Ronald, New York, 1959, Sections 35 to 42.
2. STANFORD, E. G. and TAYLOR, H. W. *Metallurgia, Manchr.* 1954, **50,** 79.
3. HOCHSCHILD, R. *Nondestruct. Test.* 1954, **12,** No. 3, 35.
4. HOCHSCHILD, R. *Nondestruct. Test.* 1954, **12,** No. 5, 31.
5. LIBBY, H. L. *Nondestruct. Test.* 1956, **14,** No. 6, 12.

REFERENCES

6. ALLEN, J. W. and OLIVER, R. B. *Nondestruct. Test.* 1957, **15,** 104.

7. McCLURG, G. O. *Nondestruct. Test.* 1957, **15,** 116.

8. McMASTER, R. C. *Nondestructive Testing Handbook*, Vol. II, Ronald, New York, 1959, Section 40.

9. BLITZ, J. *Elements of Acoustics*, Butterworths, London, 1964.

10. HOCHSCHILD, R. *Nondestruct. Test.* 1960, **18,** 323.

11. McMASTER, R. C. *Nondestructive Testing Handbook*, Vol. II, Ronald, New York, 1959, Sections 40 to 42.

12. McMASTER, R. C. and SMITH, G. H. *Mater. Eval.* 1967, **25,** 153.

3

MAGNETIC PARTICLE TESTING—PROCEDURE

3.1. BASIC PRINCIPLES

MAGNETIC particle testing is a method of finding surface and near-surface defects in any steel or iron sample capable of being magnetized. When magnetizing a component, a flux flow is generated in it and any interruption of this flow can be detected.

The usual practice consists of magnetizing the test object and then applying finely divided particles of magnetic iron oxide (Fe_3O_4) or iron filings, which are attracted to the surface at the points where cracks or other flaws cause leakage fields. A crack will have a North pole on one side and a South pole on the other. *Figures 3.1* and *3.2* illustrate this phenomenon for a straight bar and a ring.

It is essential that the flux path crosses the flaw and ideally it should be at right-angles to it. Fortunately, with suitable flux strength, defects orientated by as much as 50 degrees will show up and any object can be tested completely, provided at least two tests are made. The flux direction in the object for the second test should be at right-angles to the flux direction for the first.

To ensure an adequate test on a component, it is axiomatic to consider the following factors:

(a) The shape of the component.
(b) The dimensions of the component.
(c) The magnetic permeability.
(d) Surface finish.
(e) Defects and their orientation.
(f) A suitable flux direction.
(g) A suitable flux strength.
(h) A suitable testing stage during manufacture.

Unless due consideration is given to all these factors, the test is unreliable, although it may reveal some defects, and it is very possible that serious defects may not be revealed.

Because every component differs and at least two tests are required to find all defects, it is sound practice to record an agreed technique to ensure a consistent testing standard. For components having complex shapes, this technique may consist of as many as a dozen tests at varying strengths and using different methods to ensure 100 per cent coverage.

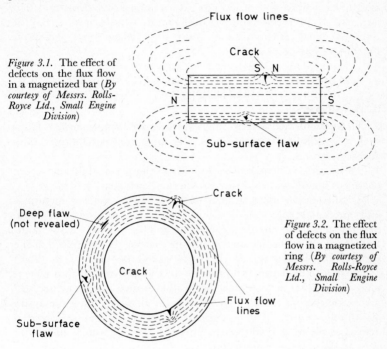

Figure 3.1. The effect of defects on the flux flow in a magnetized bar (*By courtesy of Messrs. Rolls-Royce Ltd., Small Engine Division*)

Figure 3.2. The effect of defects on the flux flow in a magnetized ring (*By courtesy of Messrs. Rolls-Royce Ltd., Small Engine Division*)

3.2. ARTEFACTS CAPABLE OF BEING TESTED

Provided that adequate equipment is available, it is feasible to test any magnetic object from a pin to the 'Queen Elizabeth'. The size of faults which can be found will depend upon surface finish and other factors, but the method is usually more than adequate for the intended use of the part, bearing in mind that only surface or near-surface defects are detected. In general engineering practice, a large proportion of components are made of steel or iron capable of being magnetized and this is fortunate, because this testing method is not expensive and it can reveal all the surface faults which matter.

It is considered essential to test all components which are subject to high stresses or fatigue and those which have been cast, welded or heat treated during fabrication. Many inspection specifications for aerospace, atomic and other critical work specifically call for this type of test. Because the method is not expensive, it is sound practice to check all steel and iron parts, since the elimination of recurring defects so found can almost certainly produce a useful saving in cost greatly exceeding that of testing. Many progressive firms test their bar material on receipt and by culling faulty lengths, ensure that the fully machined articles are not scrapped at a stage where expensive machining has been applied.

3.3. ADVANTAGES AND DISADVANTAGES OF THE METHOD

3.3.1. ADVANTAGES

(1) Magnetic particle testing can be used to find surface faults such as cracks, discontinuities such as laps, non-metallic inclusions and segregation.

(2) It can also demarcate areas where the permeability has been changed due to local cold work or thermal disturbance.

(3) Sub-surface defects such as blow holes, massive inclusions and internal cracks can be found below the surface to a depth dependent on the energizing current and the nature of the fault. Because any flux disturbance inside the material is minimized at the surface, the indication at the surface will be blurred in direct proportion to its depth. In practice, this usually means that defects more than 0·060 in (1·5 mm) below the surface (unless acute) will not be detectable. With welds where the fillet and base metal are of differing nature, it is occasionally possible to show deep-seated defects as much as 0·500 in (12·5 mm) down, although this would only apply to freak applications.

(4) To avoid contamination of the detecting media, it is always advisable to have the surface free from oil or grease, but thin even scale or foreign matter in a defect do not seriously impair detection.

(5) An overlay of paint or non-magnetic plating, such as cadmium, does not have a marked effect on the efficiency of testing, provided that it is not abnormally heavy. A cadmium plated layer is often deliberately applied to components before inspection, because the white background is helpful in testing.

(6) The test equipment is cheap, robust and can be handled by semi-skilled labour without requiring elaborate protection such as that needed for radiology.

3.3.2. DISADVANTAGES

(1) The presence of non-conducting surface coatings, such as paint, may preclude the use of contact current flow tests.

(2) The material must be capable of being magnetized, which precludes the testing of austenitic steels.

(3) Every component must be tested at least twice to ensure that flux travels in two directions at right angles and so crosses the path of longitudinal and transverse defects. Components of complex shape may need numerous tests to ensure complete coverage.

(4) After testing, each part must be demagnetized to avoid undesirable effects.

(5) The ink particles can clog fine passages and their removal is sometimes laborious.

3.4. DETECTING MEDIA

There are two broad classifications, i.e. the 'dry' and the 'wet' methods. The 'dry' method uses coarsely powdered iron filings or similar materials which are dusted on to the test object after magnetizing and then cling to any marked discontinuity. The sensitivity level is lower than for the 'wet' method but for testing rough forgings or castings, the 'dry' method is useful and will find cracks, laps and other massive faults.

The 'wet' method uses Fe_3O_4 (magnetic iron oxide) in a carefully graded powder form with fine and medium coarse particles suspended in a suitable carrier, such as paraffin. Although this kind of suspension tends to settle, it can be applied by a pipe nozzle fed through a pump from a storage tank. An agitator will obviate the tendency to settle. The sensitivity of this method is superior to that of the 'dry' method and the convenience of operation is more suited for routine testing.

The usual ink is coloured black and complies with British Standard 4069: 1966. However, red, yellow and fluorescent inks are also used. These inks are still based on Fe_3O_4 but carry dyestuffs to increase contrast. They can be extremely useful on dark surfaces or inside surfaces of restricted apertures where viewing is a problem. Their mobility, however, tends to be less than that of plain black ink, with a resultant loss of sensitivity. The fluorescent ink needs ultra-violet light ('black light') for viewing (see Chapter 4). All inks must be kept free from contamination by oil, water or cleaning fluids and should be regularly checked.

3.5. VARIOUS MANNERS OF ENERGIZING A COMPONENT

The essential requirements for any test is the application of a flux flow of adequate intensity along a known direction in the component. With any flux flow, defects can be shown basically to lie in two directions at $90°$ but for the third $90°$ direction, faults will not be observed.

3.5.1. MAGNETIC FLOW—USING MAGNET (Symbol M.F.)

The simplest possible way of magnetizing a component such as a bar is to position it across the poles of a horse-shoe magnet. The direction of flux is known and the intensity can be varied by using a strong or weak magnet, or by introducing a gap in the flux path with a thin piece of aluminium or other non-magnetic material. Defects basically transverse to the test direction will be revealed. In practice however, this method of generating magnetic flow has severe limitations and its use is confined to testing small, simple objects or to spot-checking on a suspect area of a large object (see *Figure 3.3*).

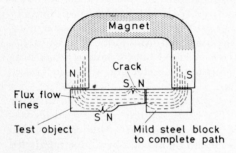

Figure 3.3. A horse-shoe magnet used to provide flux (*By courtesy of Messrs. Rolls-Royce Ltd., Small Engine Division*)

3.5.2. MAGNETIC FLOW—USING ELECTROMAGNET (Symbol M.F.)

A more refined method of producing magnetic flow is to use an electromagnet with a variable flux path, adjustable to suit the

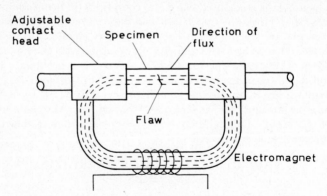

Figure 3.4. Diagram of typical electromagnet flaw detector (*By courtesy of Messrs. Rolls-Royce Ltd., Small Engine Division*)

components (see *Figure 3.4*). This method has the virtue that the energizing direct current can be varied with the aid of an ammeter reading so as to provide a wide range of ampere-turns of power and, hence, induced flux.

Provided that the pole pieces can be a snug fit on the component in question, satisfactory testing is possible for most flat or solid objects. The main snags occur with components of extreme length or varying cross-section. If the flux path length is doubled, the amount of flux flowing through the part will be halved, so that a given number of ampere-turns of energization is not a measure of the flux in the component, but only of the power applied to the flux circuit. Thus long components need a long flux path because the pole gap is greater and the increased fluxpath will be greater by twice the component length. Most commercial machines will cope successfully with parts not exceeding 12 in (300 mm) long.

Components of varying cross-section will obviously need equal variation in flux levels to give the same sensitivity of test. This means that a component with say three diameters of one, two and three inches will need ideally three tests at equivalent levels of energization. To test the largest diameter portion adequately would mean that, at the right power level, the smallest diameter portion would be saturated and not be suitable for viewing. Nevertheless, for components of simple form and reasonably regular section, this method can be most satisfactory and can give high sensitivity. It should be remembered that the defects found will be basically in a *transverse* direction to the flux flow and the test direction.

3.5.3. CONTACT CURRENT FLOW (Symbol C.E.)

By passing a current through a component, a magnetic field is formed at right angles to the current flow (see *Figure 3.5*). In

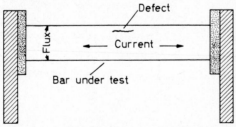

Figure 3.5. Flux flow in a bar carrying current (*By courtesy of Messrs Rolls-Royce Ltd., Small Engine Division*)

magnetic particle testing, this fact is made use of to find defects positioned basically *longitudinally* to the current path and test direction.

Low voltage is adequate, usually from 6V to 27V, which does not represent any hazard to the operator. The current must be quite high to obtain the right level of sensitivity. With high currents, it is important that the pole pieces passing the current into the work are massive and preferably faced with copper gauze to ensure that the risk of burning due to sparking is avoided. Components with pointed ends may need shaped end-fittings. This method locates defects at right angles to those found by magnetic flow so that, on an object such as a bar, the two tests applied in succession will find defects in every direction.

A particular advantage of this method is that the ammeter reading reflects precisely the flux level, whatever the component length, and it is much easier to test successive repeat components to exactly the same standard.

For objects of extreme bulk, where the positioning between machine poles would be awkward, it is quite practical to use wandercables of heavy welding type cable to feed the power from the transformer into G-clamps or other means of coupling to the work. There is naturally a voltage drop and hence a current drop with long cables but this simply means that more basic power must be fed from the transformer. The ammeter still indicates a true level of the flux generated (see *Figure 3.6*).

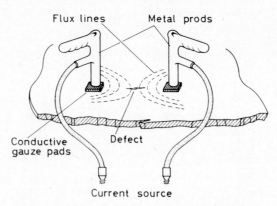

Figure 3.6. Wandercables in use with prods (*By courtesy of Messrs. Rolls-Royce Ltd., Small Engine Division*)

It should be realized that because alternating current, both recti-fied and unrectified, tends to travel on the surface of a component, the flux will stay on the surface and there will be very little flux flow inside a tube. Defects on any internal surface will not be adequately revealed unless the tube wall thickness is less than about 0·060 in (1·5 mm) when 'ghosting through' of defects will occur.

3.5.4. The Threading Bar (Symbol T.B.)

With an object such as tube, it is a useful feature that the same effect as contact current flow can be obtained by placing a threading bar through the bore and energizing it. The bar should not be of steel but of a good conductor such as copper or aluminium, and it can be hollow if the diameter is fairly large (see *Figure 3.7*). The

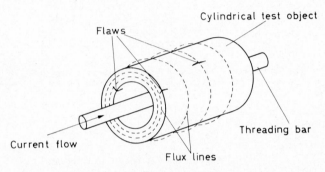

Figure 3.7. Threading bar flux flow in a small tube (*By courtesy of Messrs. Rolls-Royce Ltd., Small Engine Division*)

resultant flux field in the surrounding steel object will be at right angles to the current flow and therefore will show defects basically *longitudinal* to the current direction. A big advantage is that flux flows on the inner surface of a tube as well as on the outer surface so that internal defects can be found. There is much less risk of burning with a threading bar than with current flow but, if the perimeter is insulated, there is no risk at all.

It is important to realize that the flux strength in the object will be directly proportional to the distance of the surface from the threading bar. With a very large ring or tube, this means that, in order to detect the full perimeter at a consistent level, the tube must be partially rotated and re-tested possibly several times so that, during the multiple test cycle, the whole perimeter lies within the effective test circle round the bar. The diameter of this circle will depend directly on the energizing current (see *Figure 3.8*). The technique can be used in conjunction with wandercables to test very large components (see *Figure 3.9*).

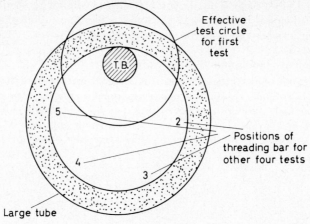

Figure 3.8. Effective field for a threading bar in a large tube (*By courtesy of Messrs. Rolls-Royce Ltd., Small Engine Division*)

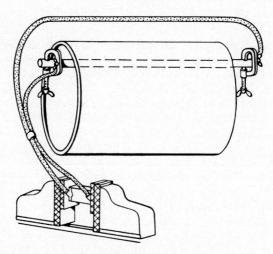

Figure 3.9. Threading bar used with prods for checking large tube (*By courtesy of Messrs. Rolls-Royce Ltd., Small Engine Division*)

3.5.5. THE COIL (Symbol Coil)

If a component is placed longitudinally within a coil carrying a current, a flux will be generated in the component, giving North and South poles at its ends (see *Figure 3.10*). To obtain the desired

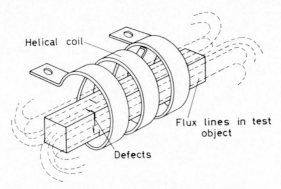

Helical coil

Flux lines in test object

Defects

Figure 3.10. Flux flow in a coil (*By courtesy of Messrs. Rolls-Royce Ltd., Small Engine Division*)

level of saturation, this coil must have several turns and be fed with a heavy current. This point is discussed in detail later (see Section 3.13). The faults revealed will be orientated basically at right angles to the flux and therefore in the transverse direction to the length of the part.

A virtue of this method of energizing is that, provided successive similar components are positioned similarly and the reading of the energizing current is the same, repeatable results are obtained. A disadvantage is that any component, in particular a short component, will have a self-demagnetizing effect dependent on the distance between the two ends. This means that a suitable current setting for one component will be different from that for another of a different length even in the same coil. Only the portion of the component within the coil is tested at full strength and repeat tests are necessary on long objects at coil width spacings.

3.5.6. INDUCED CURRENT FLOW (Symbol I.C.F.)

Large rings can be very difficult to test for radial defects because they have unfavourable shapes and do not respond well to the coil test. The induced current flow method overcomes this by the use of a special transformer (see *Figure 3.11*). This transformer has a core which can be demounted so that the ring may be threaded on it to become a single turn secondary winding. Defects will be found basically in line with the circumferential current flow, i.e. circumferentially on all surfaces.

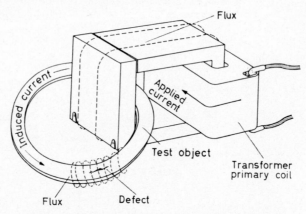

Figure 3.11. Flux flow in induced current method (*By courtesy of Messrs. Rolls-Royce Ltd., Small Engine Division*)

3.6. PRACTICAL VARIATIONS OF THE BASIC ENERGIZING PROCEDURES

The methods so far enumerated involve the use of the following:

(*a*) A *magnet*—revealing defects transverse to the test direction.

(*b*) An *electromagnet*—revealing defects transverse to the test direction.

(*c*) *Contact current flow*—revealing defects longitudinal to the test direction.

(*d*) A *threading bar*—revealing defects longitudinal to the test direction.

(*e*) A *coil*—revealing defects transverse to the coil axis.

(*f*) *Induced current flow*—revealing circumferential defects in rings.

In practice, the selection of suitable methods is limited by the basic shape of the test object and special adaptions may have to be made.

3.6.1. USE OF A SPECIAL MAGNET

For testing small suspect areas or small magnetic components mounted on non-ferrous areas, this method can be useful. A

66

horse-shoe magnet is convenient when used with suitable extension pieces of soft iron or steel, for the awkward job, and the strength can be reduced on an *ad hoc* basis by interposing, in the flux path, a thin sheet of cardboard or aluminium.

By using the magnet successively in two directions at right angles, faults in both directions can be found. It is important however to avoid local movement of the magnet on the plate or other test object, as spurious indications can be easily produced.

A magnet specially made for testing consists of a strong bar with hinged extension arms. This type is more flexible in its use but still does not solve the basic disadvantages of

(*a*) a limited test area,

(*b*) uneven flux distribution, and

(*c*) lack of control on strength (see *Figure 3.12*).

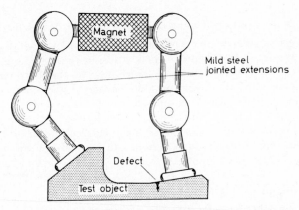

Figure 3.12. Hinged hand magnet (*By courtesy of Messrs. Rolls-Royce Ltd., Small Engine Division*)

3.6.2. USE OF AN ELECTROMAGNET WITH SPECIAL YOKE

For testing larger areas on a specific object, it is often worthwhile to make up a U-shaped block of soft iron or steel and energize this with several turns of wander cable fed from a transformer (see *Figure 3.13*). This can be made contoured to fit a specific job and can be controlled easily for the best flux level by adjusting the energizing current.

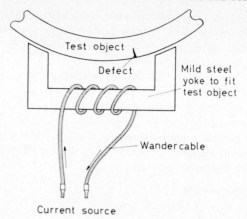

Figure 3.13. Electromagnet fed by wandercables
(*By courtesy of Messrs. Rolls-Royce Ltd., Small Engine Division*)

3.6.3. Use of Contact Current Flow—Special Variations

Testing a bar-shaped object of massive dimensions is generally easy but, for fragile or sharp ended objects, it is sometimes advantageous to make up shaped end-caps to fit the contour on the sharp ends of the component and thus avoid possible burning. Other components may be too delicate to apply end pressure when located in the testing device or physically too large to fit between poles. Wandercables can then be used with G-clamps or 'magnetic leeches' to attach to the object. Small objects such as bolts can be more speedily tested using special jigs (see Chapter 4).

3.6.4. Threading Bars—Special Types and Uses

For testing the bore of a tube, the threading bar method must be used instead of contact current flow to energize the inner surface. To ensure an even flux level, the threading bar should ideally be spaced to be located on the bore axis. This is not always convenient in practice and positioning the bar so that the tube lies on it is preferable. By increasing the threading bar diameter, a more even flux field is produced but the necessity for viewing the bore sets a practical limit to size. There is sometimes an advantage in making the threading bar non-round and a D-shaped bar can give a better flux flow, with easier viewing.

For very large diameters, where several tests are necessary with intermediate part rotation, the flux strength is always at a maximum near the threading bar. For a component with longitudinal welds (*Figure 3.14*), two or more threading bars in parallel can be used to be in line with the weld fillets. For testing critical bolt holes, a useful technique is to thread a thin wandercable or rod through the hole. Several holes can thus be tested simultaneously (see *Figure 3.15*).

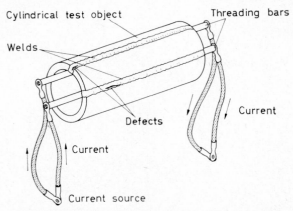

Figure 3.14. Twin threading bars used on a welded component (*By courtesy of Messrs. Rolls-Royce Ltd., Small Engine Division*)

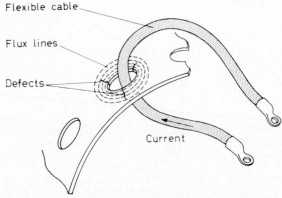

Figure 3.15. Bolt-hole checking with a flexible threading bar (*By courtesy of Messrs. Rolls-Royce Ltd., Small Engine Division*)

3.6.5. THE USE OF SPECIAL COILS

Very short objects being tested in a coil will self-demagnetize strongly. The use of slave extension pieces of mild steel, to increase the effective length, will have the twin virtues of improving the flux level and straightening the flux flow, which tends to diverge sideways at each end of a short component. Similar objects to that on test will serve satisfactorily in most cases. An example of an extension piece is shown in *Figure 3.16*. In the standard coil, it is important to test a component with its axis in line with the coil axis.

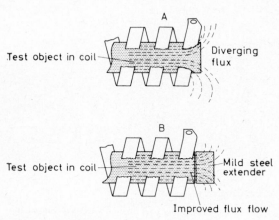

Figure 3.16. Extension piece used in a coil (*By courtesy of Messrs. Rolls-Royce Ltd., Small Engine Division*)

By using extension pieces, it is possible to produce a compound test component of favourable L/D ratio and test the component across the coil axis, thus finding defects longitudinal to the component main axis (see *Figure 3.17*). This technique is of special value when the

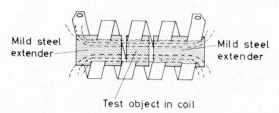

Figure 3.17. Extension pieces used in a coil to enable cross-axis check (*By courtesy of Messrs. Rolls-Royce Ltd., Small Engine Division*)

component is fragile and will not withstand the end pressure on a current flow test. In this way a component can be tested for defects

70

in all directions by using two coil tests, one being normal and the other across the axis with the use of extension pieces.

3.6.6. THE USE OF THREADING COILS (Symbol T. Coil)

Rings can be difficult to check satisfactorily where the L/D ratio is 1:1. A useful technique to find radial defects is to thread a flexible cable about three times round a portion of the ring so as to form a coil. The closed flux path occurring in the ring is of considerable help. As with the ordinary coil technique, repeat tests at coil width intervals are advisable (see *Figure 3.18*). The technique

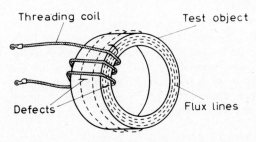

Figure 3.18. Threading coil in use (*By courtesy of Messrs. Rolls-Royce Ltd., Small Engine Division*)

can be applied to complex welded assemblies such as tubular scaffolding frames or large spoked wheels.

3.7. PROCEDURE USED FOR CHECKING A COMPONENT

3.7.1. STAGES

The following stages are necessary to ensure satisfactory detection of defects.

1. Preparation of the surface of the component (see Section 3.8).
2. Initial demagnetization (see Sections 3.7.2 and 3.11).
3. Degreasing and cleaning (see Section 3.9).
4. Application of the first test specified in the approved technique (see Section 3.12).
5. Basting of the component with the ink (see Section 3.10).
6. Viewing for defects (see Section 3.7.3. and Chapter 4).
7. Marking all defects (see Sections 3.7.4 and 3.16).
8. Demagnetizing (see Section 3.11).

9. Repetition of stages 4 to 8 as necessary for the other tests in the approved technique.
10. Removal of ink from the component (see Section 3.7.5).
11. Preparation of the component for storage (see Section 3.7.6).

3.7.2. INITIAL DEMAGNETIZATION

Components which have been machined on magnetic chucks or handled in the vicinity of any magnetic field may be magnetized either wholly or partially. It is advisable to remove this residual magnetism to avoid false indications.

3.7.3. VIEWING

The whole of the surface under test should be viewed before proceeding to the next test. Viewing of the under-surface of a component may necessitate the use of a mirror or the reversal of the component. Bores may need special lighting, and viewing of end faces may necessitate removing the component from between the contacts. Doubtful indications are often more evident if the component is allowed to drain for a few minutes.

3.7.4. MARKING OF DEFECTS

Any indications found can be marked with a grease pencil after allowing the ink to drain. For a permanent record, the indication can be lifted from the component and transferred to white paper by using adhesive tape. It is also possible to coat the component with a cold curing plastic which, when set and removed, will provide a permanent replica.

3.7.5. CLEANING

Ink particles and contrast aids (if used) can be deleterious during later assembly of the component. A paraffin oil wash supplemented, when needed, by hand brushing is usually adequate.

3.7.6. STORAGE

Because the tested component is free from oil, rusting is a possibility and a suitable corrosion preventative should be applied.

3.8. SURFACE PREPARATION BEFORE TESTING

Normally, machined or plated surfaces do not require any preliminary surface treatment other than degreasing. Parts which are to be painted should, where possible, be tested before the paint is applied.

On painted parts, where approval has been given, the paint should be removed locally, using an approved method, so as to provide adequate contact areas for the current flow tests. Other painted parts will only require degreasing unless the colour of the paint is the same as that of the particles in the ink to be used and is likely, therefore, to provide poor contrast. In the latter case an agreed contrast aid may be applied. This is normally a thin white adherent coating which is neutral to ink and readily removed after use. The application of a white emulsion paint is an alternative procedure.

Loose rust and scale should be removed from the component by an approved method to prevent contamination of the ink. Blasting methods are generally suitable if carefully controlled to avoid dimensional losses. Where a chemical method is approved, it should be chosen to avoid embrittlement of steels of high strength.

Where it is considered desirable to plug holes in a machined component (so as to avoid ink contamination of the hole or of an internal cavity leading from the hole), the plugs used should neither be magnetically affected by the ink nor obstruct viewing. Red PVC plugs, standing proud, are recommended. A hole should not be plugged if there is any possibility of cracking. This applies especially to components which have been in service.

Phosphated components will require a contrast aid or the use of a coloured ink. For contact current flow tests, the phosphate must be removed locally.

3.9. DEGREASING AND CLEANING BEFORE TESTING

The component should be thoroughly cleaned before testing, because adhering grease and dirt can mask defects and also contaminate the ink. For unassembled parts, this will normally necessitate a degreasing operation, possibly followed by a mechanical wiping operation to remove adherent solid residues. In most instances, degreasing may be carried out satisfactorily by means of white spirit or a trichloroethylene bath.

When a component is to be tested *in situ* in a structure, the following cleaning procedure is recommended:

(1) Remove the heavier soil with a clean rag or cotton waste moistened with a suitable solvent, preferably white spirit or paraffin. Liquid trichloroethylene is not suitable because the fumes can be dangerous.

(2) Using a fresh lint-free rag and fresh solvent, carry on cleaning the area to be tested until a satisfactory clean surface is obtained.

It is important that the solvent is taken to the rag or cotton waste, and not vice versa, otherwise contamination of the solvent will occur and grease and soil will be transferred back on to the component.

3.10. APPLICATION OF INK (OR MAGNETIC POWDER)

If a dry powder is used, it should be applied to the energized component with a pepper pot type applicator so as to obtain an even distribution. Tapping the component with a rubber hammer is often helpful.

For best results, the ink should be applied during energization and should cease flowing just before excitation ceases. Whether brush, ladle or hose application is used, care should be taken to avoid violent flow over the component, which could disturb faint indications. The flow should cease before the excitation is switched off.

3.11. DEMAGNETIZATION

Residual magnetization can affect the testing, cause chip adherence in service, and have an adverse affect on compasses and other delicate service equipment. It is advisable to demagnetize:

(*a*) before testing,
(*b*) between specified tests in which the flux direction changes,
(*c*) after testing.

For aerospace applications, a compass test is specified for finished parts. Electron-beam welding is an application for which residual magnetism can be a great nuisance by deflecting the beam. Care should be taken that components are demagnetized individually as bulk demagnetizing is not effective.

Some components such as rings and very short specimens are difficult to demagnetize owing to their poor L/D ratios. In such cases, the addition of a long dummy extender steel strip held in close proximity to the part will alter the shape favourably and allow satisfactory demagnetization (see *Figure 3.19*).

In all cases, the component being demagnetized should be removed clear of the flux field before switching off the current to the demagnetizer. This can be accomplished by bringing the component at least 3 feet (1 m) clear of the demagnetizer along the axis or

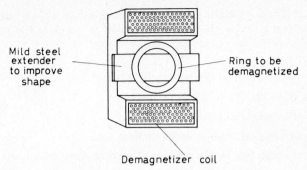

Mild steel extender to improve shape

Ring to be demagnetized

Demagnetizer coil

Figure 3.19. Extender used to facilitate demagnetization (*By courtesy of Messrs. Rolls-Royce Ltd., Small Engine Division*)

simply by bringing the component to the dead zone outside the coil (see *Figure 3.20*).

To avoid accidental re-magnetization, demagnetized components should not be left in the vicinity of the demagnetizing coil.

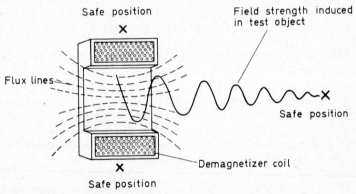

Safe position

Field strength induced in test object

Flux lines

Safe position

Demagnetizer coil

Safe position

Figure 3.20. Demagnetizer diagram showing alternative safe zones (*By courtesy of Messrs. Rolls-Royce Ltd., Small Engine Division*)

The efficiency of demagnetization should be checked by using a compass or a commercial magnetic field indicator on at least one component in a batch.

In addition to using the usual commercial demagnetizer it is possible to demagnetize by

(*a*) Heat treating the component above 715°C. This may be a requirement during the subsequent finishing of a heat-treated part.

(*b*) Using a standard machine coil energized by a.c. and reducing the current in stages not greater than 50 per cent of the previous value down to the minimum possible value.

3.12. SETTING A TECHNIQUE

The test results can be strongly affected by:

(*a*) the magnetizing operation used,

(*b*) the test values of current or flux,

(*c*) the jigs or fixtures, such as threading bars.

For any component it is therefore advisable to devise and write down an agreed technique listing the operations required with all details of the directions of tests and jigs needed. The use of a set technique will ensure consistent testing to the same standard and can avoid inconsistent results being obtained by different operators, with consequent dispute between supplier and user.

It is often found that, due to lack of understanding of suitable values, parts are consistently under-tested or over-tested. Over-testing may appear to be harmless but it can cause acceptable material to be rejected, with heavy cost thus involved.

In devising a technique, the geometrical shape of the object is of major importance and *Table 3.1* gives guidance. It should be noted that the basic technique ignores critical testing of end faces of bar stock and the like. If these are important, a supplementary test should also be applied.

As would be expected, many components do not have simple geometrical shapes but have what can be considered to be combinations of simple shapes. For instance, a flat gear with a long centre shaft would be tested as a disc for the gear portion and a bar for the shaft. Recommended test values for critical work are given in Section 3.13.

When components such as a stepped bar vary considerably in diameter, it may mean that more than one test in the same direction and manner is needed to inspect adequately all the different areas. When the high level test values recommended are used, the steel saturation value is about 40 per cent so that it would be reasonable to check a part varying from 2 in (50 mm) diameter to 1 in (25 mm) diameter at the 2 in (50 mm) recommended value without causing over-saturation. However, if the stepped range of diameters were say 4 in to 1 in, testing at the 4 in value would over-saturate the one-inch area and two tests at different values would be needed. The lower value tests should be made first.

Table 3.1a. Suitable tests for material or component of specific geometrical shape

Component shape	Basic tests giving commonly specified degrees of coverage see Table 3.1b	Supplementary magnetizing operations to be added when full coverage of ends or minor faces is essential
Long bar	1 & 4 or 1 & 3	1T & 1ST or 4T & 4ST or 1T & 4T
Short bar	1 & 4 or 1, 1T & 1ST or 4, 4T & 4ST	As for long bar
Helix (spaced winding) e.g. spring	1 & 2 or 1, 4T & 4ST	As for long bar if practicable
Long Tube (see Note 6)	2 & 4 or 2 & 3	1T & 1ST
Short Tube (see Note 6)	2 & 4 or 2, 1T & 1ST or 4, 4T & 4ST	As for long tube
Long closed end tube	2 & 4 or 2 & 3	Closed end: As for long bar Open end: 1T
Short closed end tube	As for short tube	Closed end: As for long bar Open end: 1T & 1ST
Large tube	6 & 7T or 7 & 7T	6, 7T & 7ST
Small ring	2, 1T & 1ST or 1T, 1ST, 4T & 4ST or 1 & 5	None
Large ring	6, 7T or 2 (Repeat at stated intervals with T.B. close to inner surface) & 7T or 5 & 6	None
Large plate	7 & 7T	On all minor faces 7 & 7T
Small plate	1 & 1T or 4 & 4T or 1 & 4 or 1 & 3	On all minor faces 1 & 4
Large disc	7 & 7T	On all minor faces 7 & 7T
Small disc	1T & 1ST or 4T & 4ST or 1T & 4T	By performing any two alternatives, the minor faces will be covered
Large sphere	7 & 7T	None
Small sphere	1, 1T & 1ST or 4 & 4T or 3 & 3T (Use extenders both tests)	None

Table 3.1b. List of possible tests

Reference	Excitation method	Direction applied	Sensitive to defects in direction of			Descriptive symbols to be used
			Major axis	Transverse	Short transverse	
1	CF	Major axis	✓	X	X	1/CF/ current value
IT	CF	Transverse to major axis	X	✓	✓	IT/CF/ ,, ,,
1ST	CF	Short transverse to major axis (on round object at 90° to 1T)	X	✓	✓	1ST/CF/ ,, ,,
IT 60° (120°)	CF	At the quoted degrees angle to 1T on a round object	X	✓	✓	IT60°/CF/ ,, ,,
1X	CF	Special direction as detailed	?	?	?	1X/CF/ ,, ,, + direction detail
2	TB	Through main bore	✓ including bore defects	X	X	2/TB/ ,, ,,
2T	TB	Through transverse subsidiary bore	X	✓	✓	2T/TB/ ,, ,,
2ST	TB	Through short transverse subsidiary bore	X	✓	✓	2ST/TB/ ,, ,,
2X	TB	Through nominated subsidiary hole	?	?	?	2X/TB/ ,, ,, + details
3	Coil	Major axis along coil bore	X	✓	✓	3/Coil/ ,, ,,
3T	Coil	Transverse to coil bore (only applicable where extenders are used)	✓	X	✓	3T/Coil/ ,, ,, + coil details
4	MF	Major axis	X	✓	✓	4/MF/ ,, ,,

78

4T	MF	Transverse to major axis	√	X	√	4T/MF/current value
4ST	MF	Short transverse to major axis	√	√	X	4ST/MF/ " " " "
		On non-round objects (on round object at 90° to 4T)				
4T 60°	MF	As the quoted degrees angle to 4T		?	?	4T60°/MF/ " " " "
4X	MF	Special directions as nominated		?	?	4X/MF/ " " " "
5	1.CF	Major axis	circumferential only			5/1CF/ " " " "
6	T. Coil	Round ring or bar section:				6/T. Coil/ " " " "
		Ring	√	√*	√	
		Bar	X	√	X**	
7	CF (Prods)	........As Test 1				6/CF (Prods) current value
7T	CF (Prods)	........As Test 1T				7T/CF (Prods) " " " "
7ST	CF (Prods)	........As Test 1ST				7ST/CF(Prods) " " " "
7X	CF (Prods)	Special direction nominated	?	?	?	7X/CF(Prods) " " " "

* Circumferential
** Radial

These tests are listed in the preferred order in *Table 3.1a*.
The technique selected must include tests to cover defects in all three possible directions.
The alternatives allow for this requirement.

79

3.13. TEST VALUES

3.13.1. SENSITIVITY LEVEL

If the defects sought are massive cracks, very low levels of magnetization will give sufficient build-up, but to find minute cracks or subsurface irregularities such as manganese sulphide inclusions, the flux level should be high enough to give the maximum build-up. Theoretically, a level just below saturation would give the most sensitive result but this is impractical owing to the non-regular shapes encountered and, in practice, lower levels are quite adequate. Most constructional steels have permeabilities which are sufficiently similar to enable a common value of flux level to be used for them all. Freak materials such as transformer laminations are an exception.

The sensitivity level can be made quite low for massive cracks but for critical testing, such as in aerospace work, an accepted standard is about 40 per cent of saturation. This is high enough to find small defects and is not too difficult to obtain. The sensitivity at this level is sufficient to cause the indication on a surface crack to be approximately 200 times the actual crack width. Subsurface defects will also show, although the indications become diffuse with increasing depth. Large defects 0·100 in (2·5 mm) below the surface can always be found and, up to about 0·060 in (1·5 mm) deep, all defects of a serious nature can be detected. With compromise settings to cater for stepped diameter components, the lowest flux level is still reasonable and the highest level can be kept below saturation, which will produce excessive background.

The values quoted below are based on a 40 per cent saturation level and in many cases are similar to those quoted in British Standards and some U.S.A. standards.

3.13.2. MAGNETIC FLUX

With permanent magnets and electromagnets, the flux is generated directly into the part. Variables affecting the flux include:

(a) the cross-section of the part,

(b) the length of the part,

(c) the total length of the flux path including the machine poles.

The ammeter reading on the d.c. exciting current only indicates the number of ampere-turns passing through the yoke and *not the requirement for the component*. Measurements can be made with oersted meters but owing to the nature of the variables, it is much more practical to find the saturation level of the part by increasing the flux level until background is formed, and using a lower value of about 40 per cent of this.

A recommended test level is one below but not less than one-third of saturation level. For an electromagnet, this can be expressed in terms of an ammeter reading if other conditions are kept constant.

3.13.3. CURRENT FLOW

With the usual contact method of current flow, the main variable is the perimeter of the component. The ammeter reading indicates the excitation strength in the part. It should be noted, however, as discussed in Chapter 4, that it is the *peak current which matters* and this may be considerably higher than that shown on the meter which normally reads r.m.s. values.

Unless the ammeter is specially marked, the ratio of peak to r.m.s. current is as follows:

Source	Meter Reading A	Actual Peak Current A
D.C. battery	1	1
A.C. mains	1	$\sqrt{2}$
D.C. rectified half wave	1	$2\sqrt{2}$
D.C. rectified full wave	1	$\sqrt{2}$

The peak current values are recommended and they should be divided by the appropriate ratio shown above to obtain indicated ammeter readings for the various power sources. *A recommended test level for critical work is 230A (peak) per inch (9A/mm)* of perimeter, ignoring minor variations of contour. For basically round components this can be expressed as 720A (peak) per inch (28A/mm) of diameter. When current flow is used with prods on a large surface, different conditions apply as discussed in Section 3.17.

3.13.4. THREADING BAR

With the usual method for which a bar passes through the bore of a tube, the current suggested in Section 3.13.3 can be applied if the bore is concentric with the bar. In many instances, particularly for large tubes or rings, the diameter is too large to enable the suggested value to be applied. For instance a 12 in (300 mm) o.d. tube would theoretically need $12 \times 720A$ (peak) = 8,640A. However, if the threading bar is deliberately placed off-centre, the portion near the bar can be tested with much lower currents and, by repeat testing after partial rotation of the tube, the whole perimeter can be examined by ensuring that each portion is within the test area.

The effective test area will be that portion of the tube within a circle centred on the bar, provided that the diameter of the circle is consistent with the available current in accordance with Section 3.13.3. For example, an available current of 720A × 3 would be used for testing a 3 in (75 mm) disc to the suggested standard.

When weld lines and other specific zones only are to be examined, a single test with the bar as near as possible to the test area would suffice. For checking bolt holes with a cable or bar through the hole, quite low values would be suitable to check the zone round the hole. For example, 720A would be sufficient to check a 1 in (25 mm) bolt hole to the suggested standard.

3.13.5. Coil

Whatever shape coil is used, the ideal is to produce a magnetic field strength which is below but not less than say 30 per cent of the saturation level for the component. This is not easy to specify because the factors involved are complex and include the following:

(a) Current applied to coil ⎫
(b) Number of turns in coil ⎬ Number of ampere turns.

(c) Fill factor of coil when testing a component (see Section 2.2.2)

(d) Length over diameter ratio of component.

(e) Coil shape.

The number of ampere turns can be easily calculated and values found suitable for a given coil with a specified component can obviously be repeated to provide consistent testing conditions.

For efficiency, the fill factor should be high and the smallest convenient coil size is recommended. However, an improvement in efficiency is obtained by keeping the component near the coil perimeter and not at the centre. With an 8 in diameter coil and a 1 in diameter component, this can produce ten times greater efficiency.

The length over diameter ratio of the component controls its self-demagnetizing effect which is reasonably consistent and low for a ratio over 15:1 but becomes massive when the ratio is below 5:1 and excessive when it is below 3:1. Extenders can be used to minimize this feature, see Section 3.14, para. (6).

To obtain a suitable test value, a very practical method, where sufficient current is available, is to increase deliberately the exciting current with the component in position until saturation is obtained. This is evidenced by furring of the component and the suggested standard would then be 40 per cent of the saturation value.

3.14. GENERAL PRECAUTIONS

For rigid coils, a workable value for any component can be derived from the formula below, provided that the component is placed near the coil perimeter and has a cross-sectional area not more than 10 per cent of the coil cross-sectional area:

$$NA = \frac{32,000D}{L}$$

where N = effective number of turns of the coils (usually 4 or 5), A = current (peak) in amperes, D = diameter of the part in inches (or millimetres) and L = length of the part in inches (or millimetres).

A suitable value for a threading coil on a ring or a component with an L/D ratio over 15 is:

$$A = \frac{360}{CT}$$

where C = coil diameter in inches and T = number of coil turns.

In metric form where C = diameter in millimetres, the formula becomes:

$$A = \frac{9,000}{CT}$$

3.14. GENERAL PRECAUTIONS

The following precautions should be taken in magnetic particle testing:

(1) Keep the magnetic ink free from water or oil, which can cause coagulation and loss of sensitivity. A 100 ml Crowsfoot glass measure is a useful adjunct. At regular intervals, a representative sample of the ink should be placed in the measure and allowed to settle. The amount of solid Fe_3O_4 settled after one hour should be comparable with that for new ink and the concentration should generally not be less than 1·25 per cent. The ink fluid should be clear; any brown coloration or muddiness indicates the presence of contaminants.

Because the magnetic ink tends to settle on standing, stir up the machine tank after periods of disuse and verify that feed pipes are not blocked.

(3) Do not mix brands of ink unless compatibility has been established.

(4) When using magnetic flux circuits, remember that copper gauze or other non-magnetic materials will reduce the magnetic flux level. Their uses must thus be carefully controlled.

(5) The values of current used in testing are sometimes massive and excitation times of over 3 seconds should be avoided since heating of the work can occur.

(6) Small parts with L/D ratios below 5:1 are difficult to test in coils since the demagnetization effect is strong and the magnetic flux path in the part may not be truly axial. It is thus sound practice for short components to use 'extenders' of mild steel in contact with the ends. Further components similar to those under test can be used for the purpose. The flux flow is then straighter and the required current lower.

3.15. ASSESSING INDICATIONS

3.15.1. CRACKS

The most dangerous defect to be found is a crack, which is normally so deleterious that rejection of the part is inevitable. Cracks are easy to reveal, whether clean or filled with foreign matter, because they have a marked effect on the flux flow. Their positions and natures offer clues to their cause thus enabling correction procedures to be adopted on future batches. Heat treatment cracks, which are normally long and narrow, occur at a marked change of section or at sharp contours such as the edge of a hole or root of a thread. Grinding cracks occur initially in a direction at right angles to the grinding and, if severe, may form a 'crazy paving' network of surface cracks. Shrinkage cracks may be present in castings at marked changes in section.

Laps and folds in wrought material can usually be recognized as they tend to be shorter and wider than cracks and do not follow the grain flow.

3.15.2. INCLUSIONS

Inclusions due to manganese sulphide streaks are less dangerous and always occur, in line with the grain flow of the part, as straight continuous or discontinuous lines. Their effect on the part is small unless they are very massive or they occur at a critical area of the component. Inclusions due to oxides show angular outlines along the indications and are less tolerable.

3.15.3. DIFFUSE INDICATIONS

Any indication of a diffuse nature is normally indicative that the defect is subsurface and the degree of spread increases with depth.

3.15.4. Spurious Indications

Spurious indications can be obtained by magnetic writing caused by a magnetized component being allowed to rub against a hard pointed surface such as that of another component. They can also be caused by local cold work due to some machining fault. They are indefinite in nature and will disappear if the part is carefully de-magnetized and re-tested.

3.15.5. Local Diffuse Indications

Local areas on a component showing build-up may be caused by 'ghosting through' of interior changes of sections e.g. those found with a bore spline. This effect may sometimes be due to a local change in permeability due to some factor in the previous history of the part, such as local spot heating or local heavy cold working. Also surface decarburization or carburization can show patches of local diffuse indications.

3.16. ACCEPTANCE STANDARDS

In assessing whether a component with indications is (*a*) service-able, (*b*) capable of being rectified or (*c*) a reject, it is of prime importance to ensure that the test conditions and, hence the build-up on the indication, are precisely similar to that used for the standard. The use of too high current values will give more marked indications and, if not controlled during assessment, can lead to rejection of satisfactory parts.

For comparison purposes, it is possible to preserve indications in several ways, as follows:

(*a*) Allowing the component to drain and coating it with clear lacquer.

(*b*) Covering the indications with clear sticky tape.

(*c*) Picking up the indications with clear tape and re-positioning them on white paper.

(*d*) Covering the component with cold curing plastic such as PVC Paste DD501 (S. Dugdale), and cutting off the replica when set.

(*e*) Using a photographic record, preferably under identical lighting conditions to that used for viewing.

3.17. PROD TESTING (Symbol C.F. (Prods))

Prod testing is a term applied to the current flow testing of large objects when power is fed to wandercables which are used with

85

contact prods attached (see *Figure 3.6*). This method can be used to test very large areas such as the hull of a ship, but to ensure consistent results the vagaries of the magnetic field produced should be realized and suitable test conditions established.

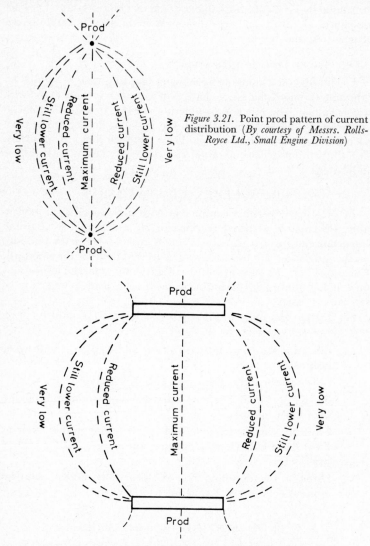

Figure 3.21. Point prod pattern of current distribution (*By courtesy of Messrs. Rolls-Royce Ltd., Small Engine Division*)

Figure 3.22. Wide prod pattern of current distribution (*By courtesy of Messrs. Rolls-Royce Ltd., Small Engine Division*)

It is not advisable to use point prods because the current intensity is very high locally and burning can take place. The current intensity between the two prods will be at a maximum on the axis and will progressively reduce according to the relative resistance of the current path, which must be longer as the distance from the axis increases. This is illustrated in *Figures 3.21* and *3.22.*

In a test ellipse such as that illustrated in *Figure 3.21*, it can be shown that on the *midpoint* of the *centre line* between the *point prods*, the current per linear unit of width on the test surface will be:

$$\text{Current density (A/linear unit)} = \frac{0\cdot637\,T}{d}$$

where $T =$ total current applied and $d =$ Prod gap in the units chosen. For example, 2,000 A applied at a gap of 10 in (254 mm) would have a maximum centre current of 1,274A/linear inch (25 mm).

In order to select suitable conditions for prod testing, it is preferable to choose a suitable value which will not saturate the steel at the centre line. With most steels, this would occur at about 575A peak/linear inch (22·5A/mm) if the current travels mainly near the test surface.

Table 3.2. Centre current using point prods

Required Max. 5,75A/linear inch   22·5A/linear mm.
Min. 2,30A/linear inch   9A/linear mm.

Prod gap		Total current applied A (*peak*)							
Inch	mm	1,000		2,000		3,000		4,000	
		linear inch	linear mm	linear inch	linear mm	linear inch	linear mm	linear inch	linear mm
1	25	*636*	24·9						
2	51	318	12·4	*636*	24·9				
3	76	212x	8·29	424	16·6	*636*	*24·9*		
4	102	x	x	318	12·4	477	18·7	*636*	*24·9*
5	127			254	9·94	382	15·0	509	19·9
6	152			212x	8·29	318	12·4	424	16·6
8	216			x	x	239	9·35	318	12·4
10	254					191x	7·47x	254	9·94
12	305					x	x	212x	9·28x

x = Below suggested standard.   Italics = above required standard.

87

A pulsating current (which is normally used) will travel near the surface but it should be borne in mind that a plate of thickness more than about 0·060 in (1·5 mm) would not be tested adequately on the under-surface and, ideally, it should be re-tested.

Similarly, it is desirable to select a minimum acceptable test value which will then define the effective width of a test ellipse. For critical work, a value of 230A(peak)/linear inch (9A/mm) as is used for ordinary contact current flow is suggested. *Table 3.2* shows some typical values using these criteria and illustrates the limited acceptable test condition so obtained. Although point prods are often used, they are deprecated because the test conditions must be restricted and the risk of burning and saturation near the prods is high.

If point prods are used, the *effective width* of the test ellipse using the same minimum acceptance value of 230A(peak)/linear inch (9A/mm) can be derived from the formula:

$$\text{Test width } (x - x) = 2 \left[\frac{d}{2\pi} \frac{T}{230} - \left(\frac{d}{2}\right)^2 \right]^{\frac{1}{2}}$$

$$= 2 \left[\frac{dT}{1445} - \left(\frac{d}{4}\right)^2 \right]^{\frac{1}{2}}$$

where d = prod gap in units chosen, $x - x$ = maximum width, and T = Total applied current, A (peak).

Table 3.3 shows typical test values.

If *wide prods* are used, the current distribution is much improved. Thus the prods should be as wide as possible to reduce burning risks and to increase the test ellipse size. The *current density midway on centre line of the prods* will vary as below:

$$\text{Current density (A/linear unit)} = \left[\frac{2}{\pi} \tan^{-1}\left(\frac{W}{d}\right) \right] T_u$$

where W = prod width in the chosen unit, d = prod gap in the chosen unit, T_u = current per unit width of prod, A (peak), and $2 \tan^{-1} (W/d)$ = the angle subtended at the centre point by a prod (quoted in radians).

This can be rewritten as follows:

$$\text{Current density (A/linear unit)} = \left[0·637 \tan^{-1}\left(\frac{W}{d}\right) \right] T_u$$

Table 3.3. Maximum width of test ellipse with point prods

Prod gap		Total current applied A (*peak*)							
Inch	mm	1,000		2,000		3,000		4,000	
		inch	mm	inch	mm	inch	mm	inch	mm
1	25	*1·33*	*33·8*	*2·13*	*54·1*	*2·70*	*68·6*	*3·17*	*80·5*
2	51	1·24	3·1	*2·66*	*67·6*	*3·55*	*90·2*	*4·26*	*108·2*
3	76	x		2·76	70·1	*3·99*	*101·3*	*4·92*	*125·0*
4	102			2·48	63·0	4·15	105·4	*5·32*	*135·1*
5	127			1·64	41·7	4·06	103·1	5·51	140·0
6	152			x		3·72	95·5	5·52	140·2
7	178					3·02	76·7	5·34	135·6
8	216					1·56	39·6	4·96	126·0
9	229					x		4·32	109·7
10	254							3·28	83·3
12	305							x	

x = Below standard.   The values in italics would give over saturation on the centre line.

The angle can be determined trigonometrically. *Table 3.4* shows typical values derived from this formula. *Figure 3.23* shows the comparison in current distribution using various prod widths.

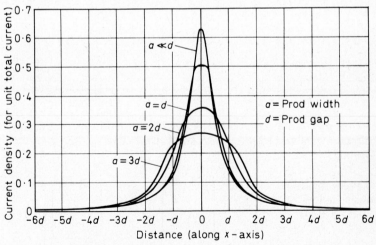

Figure 3.23. Graph showing distribution of current with prods of varying width
(*By courtesy of Messrs. Rolls-Royce Ltd., Small Engine Division*)

Table 3.4. Centre axis current when using prods of finite width

Formula: Current in A/linear unit = 0·637 × T_u × radian of half the angle at centre subtended by prod.

Ratio Prod gap/ Prod width	Subtended angle	Subtended angle, radians	Centre current in A/linear unit for an applied current per unit of prod width					
			500 inch	19·7 mm	1,000 inch	39·4 mm	2,000 inch	78·7 mm
9:1	12° 41′	0·2214					141x	5·55x
8:1	14° 15′	0·2487					158x	6·22x
7:1	16° 16′	0·2839					181x	7·12x
6:1	18° 56′	0·3304			105x	4·13x	210x	8·27x
5:1	22° 37′	0·3947			126x	4·96x	251	9·88
4:1	28° 4′	0·4899			156x	6·14x	312	12·3
3:1	36° 52′	0·6434	102x	4·01x	205x	8·07	410	16·1
2·5:1	43° 36′	0·7610	121x	4·76x	242	9·53	485	19·1
2:1	53° 8′	0·9273	148x	5·83x	295	11·6	590	23·2
1·5:1	67° 22′	1·1758	187x	7·36x	374	14·7	749	29·5
1:1	90° 0′	1·5708	250	9·84	500	19·7	1000	39·4
0·5:1	126° 52′	2·2142	352	13·86	705	27·7	1410	55·5
0·25:1	151° 56′	2·6517	422	16·6	844·1	33·2		

x = Below suggested standard.   Italics = above suggested standard.

For the same criteria of 230A(peak)/linear inch (9A/mm) minimum, the test ellipse maximum width can be derived from the formula

Current density at selected point (A/linear unit)

$$= \frac{1}{\pi}\left[\tan^{-1}\left(\frac{P+\dfrac{W}{2}}{\dfrac{d}{2}}\right) - \tan^{-1}\left(\frac{P-\dfrac{W}{2}}{\dfrac{d}{2}}\right)\right] T_u$$

where P = distance from centre line to selected point.

90

By extrapolation, a value of 230A/linear inch (9A/mm) is obtained at a point P where the radian of the subjected angle by a prod satisfies the equation below:

$$230 = \frac{\text{radian } P_o \times T_u}{\pi}.$$

or, $\qquad 723 = \text{radian } P_o \times T_u$

x — x will equal $2\times$ distance of P from centre line (which can be determined trigonometrically).

Table 3.5. Ratio of width of test ellipse (in proportion to prod width) for various prod gap/prod width ratios

Current A/ Unit Prod	Subtended angle	Subtended angle, radians	6:1	5:1	4:1	3:1	2:1	1:1	0·5:1
250/inch 9·8/mm	166°	2·892							
500/inch 19·7/mm	83°	1·446					x	0·50	0·93
750/inch	55°	0·964					x	1·17	1·22
1,000/inch 394/mm	41°	0·723				x	1·24	1·49	1·38*
1,250/inch	33°	0·578			x	1·10	1·76	1·77	1·51*
1,500/inch	28°	0·482		x	0·48	1·87	2·15	1·96	1·62*
1,750/inch	24°	0·413		x	1·72	2·39	2·47	2·14*	1·72*
2,000/inch 78·7/mm	21°	0·362	x	1·58	2·52	2·80	2·76*	2·30*	1·82

For example, if 3 in wide prods were used at 6 in spacing (2/1 ratio) with an applied current of 3,000A (1,000A per prod inch), then the ellipse would have a maximum width of 3 in $\times$ 1·24 = 3·72 inches to obtain the required standard.

* With these settings, the centre current will be higher than recommended.

Having selected suitable test conditions and determined an acceptable test ellipse, the search pattern to ensure full coverage of an area should be such that no areas are tested below the desired standard. *Figure 3.24* shows suitable coverage.

Note The values obtained and quoted in the tables show that the conditions needed to obtain the suggested standards for prod testing are restrictive. It is appreciated that for many applications, a much lower standard could be accepted and satisfactory tests at settings outside these recommendations could be made. In particular, sheet metal thinner than 0·082 in (2 mm) would have a higher surface current and thus allow the safe use of wider test ellipses.

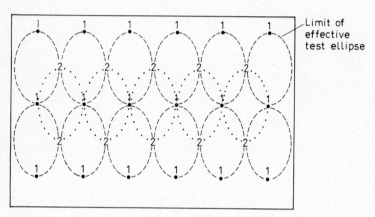

Test four times. 1. At 1 centres for N–S flaws
* 2. At 2 centres for N–S flaws
  3. At 90° to test 1 for E–W flaws
* 4. At 90° to test 2 for E–W flaws

* These tests are offset from the previous test by half the test ellipse sideways and by half the prod spacing downwards.

Figure 3.24. Suggested test patterns for prod testing (*By courtesy of Messrs. Rolls-Royce Ltd., Small Engine Division*)

REFERENCES

Relevant National Standards

Several recently issued B.S. Specifications deal with Magnetic Particle Testing and may be of interest.

B.S. 3683. Glossary of Terms used in Non-destructive Testing. Part 2: 1963. Magnetic Flaw Detection.

B.S. 3889. Methods for Non-destructive Testing of Pipes and Tubes. Part 4A: 1965. Magnetic Particle Flaw Detection: Ferrous Pipes and Tubes.

B.S. 4069: 1966. Magnetic Flaw Detection Inks and Powders.

B.S. 4080: 1966. Methods for Non-Destructive Testing of Steel Castings.

B.S. 4124. Non-Destructive Testing of Steel Forgings. Part 2: 1968. Magnetic Flaw Detection.

4

MAGNETIC PARTICLE TESTING EQUIPMENT

4.1. INTRODUCTION

IT IS emphasized that magnetic particle testing is an important process in the production of steel components which *are not checked at any subsequent manufacturing stage.* The provision of adequate equipment for the use of a reliable operator will more than justify the initial cost and will ensure that the tests are correctly carried out. In practice, power operated machine poles and other devices to mechanize operations have been found more than to repay initial costs by increasing the output per operator, who is the most expensive charge on any process.

4.2. SELECTION OF EQUIPMENT

In most engineering works, it is usual for a range of items of varying shapes and sizes to be handled. Thus the criteria for selecting equipment should take note of the smallest and the largest items, as well as freak shapes. Where the components are specialized and consist only of, say, tubes of various sizes, the possibility of improving efficiency and saving money by using specialized equipment should be considered.

Typical items of suitable equipment are listed below and discussed in detail later in this Chapter.

4.2.1. REQUIREMENTS FOR SMALL FIXED INSTALLATIONS

Equipment suitable for small items such as shafts, gears, etc., up to say 2 ft (600 mm) long, in small quantities includes:

(*a*) A flaw detector with 3,000A a.c. output (on short-circuiting the poles), with, say, a 3 ft (1 m) pole gap.

(*b*) A source of magnetic flux to provide a field of about 800 oersteds (64 kA/m). This can be combined with (*a*).

(*c*) Alternatively to (*b*), a 4 in (100 mm) diameter coil with provision for attachment to (*a*).

(*d*) A demagnetizer with, say, an 8 in × 6 in (200 mm × 150 mm) opening.

(e) Threading bars of $\frac{1}{2}$ in (12·5 mm) diameter by 2 ft (0·600 m) long and 1 in (25 mm) diameter by 3 ft (1 m) long.

(f) An ink source and means of application, such as a ladle.

(g) Checking equipment for machine and ink. This consists basically of a test-piece and a crowsfoot receiver.

(h) Means of viewing such as a 100 watt pearl bulb in a flexible mounting.

(i) A grease pencil to mark defects.

(j) Special jigs, as required.

4.2.2. Requirements for Medium Size Installations

For bulkier work up to, say, 5 ft (1·5 mm) long and 1 ft (0·3 m) in diameter, more power will be needed to maintain the desired flux levels, and the following are suitable:

(a) A flaw detector with 5,000A a.c. output on short-circuit. Powered heads to speed output are justified.

(b) A magnetic flux source to provide a field of up to about 1,500 oersteds (120 kA/m).

(c) Alternatively to (b), three coils of say 4 in (100 mm), 8 in (200 mm) and 12 in (300 mm) diameters.

(d) A demagnetizer of about 14 in × 10 in (360 mm × 250 mm) opening.

(e) Threading bars, as for small installations, plus hollow threading bars of 2 in (50 mm) diameter by 4 ft (1·2 m) long and 3 in (75 mm) diameter by 5 ft (1·5 m) long.

(f) An inkwell fitted with an agitator and a pump to feed a delivery hose and nozzle.

(g) Checking equipment as in Section 4.2.1.

(h) Means of viewing, e.g. 5 ft (1·5 m) fluorescent tubes.

(i) A marking pencil.

(j) Support blocks to suit the work envisaged.

(k) Special jigs, as required.

4.2.3. Requirements for Large Equipment

It is not feasible to detail all one's requirements; the more important of these are:

(a) The flaw detector, whether a.c. or d.c., should have an output in amperes (peak), of not less than 115 per in (5 per mm) perimeter of the largest object and, if feasible, not less than 230 per in (10 per mm).

4.2. SELECTION OF EQUIPMENT

The pole gap should accommodate the longest component to be tested which can be 20 feet (6 m) or longer. For heavy work it is advisable to have rotating pole pieces and powered movements of the heads are essential.

(b) A flux source to produce saturation of the most massive item to be tested, with a minimum field of 1,500 oersteds (120 kA/m).

(c) Alternatively to (b), coils as for medium equipment plus 16 in (400 mm), 20 in (500 mm) or even larger coils, as needed. For efficiency, there should be about 4 in (100 mm) difference between each diameter throughout the range.

(d) A demagnetizer to suit the largest item. Where the sizes are extreme, provision should be made to demagnetize by current reduction or a wiper demagnetizer should be used as an auxiliary.

(e) Threading bars, as for medium size installations, plus any special jigs to suit the very large items under test.

(f) Ink source as in Section 4.2.2.

(g) Checking equipment as in Section 4.2.2.

(h) Viewing equipment, such as twin fluorescent tubes plus a hand fluorescent lamp.

(i) A marking pencil.

(j) Support blocks.

(k) Lifting equipment to facilitate loading.

(l) Wandercables and Gee-clamps.

(k) Special jigs as required.

4.2.4. PORTABLE EQUIPMENT

For site testing, the equipment must be capable of being man-handled up ladders and to be located remote from mains supply. Where mains current is available, the equipment can consist basically of:

(a) A mains operated lightweight transformer in a robust case. This should have an output of 1,500A a.c.

(b) Two wandercables about 4 ft (1·2 m) long.

(c) Gee-clamps.

(d) A split coil.

(e) A squirt can of magnetic ink or a can of ink with a brush.

(f) An extra long wandercable, say 6 ft (1·8 m) long to be used as a coil or a demagnetizer. Where mains power is not available, powerful magnets can be used, although it will be laborious to check large areas. Alternatively, a large car or aircraft battery of 6 or 12 volts will give useful current outputs up to about 500 amperes and can be used with wandercables to check quite large components for cracks, although the lower saturation levels feasible would not permit detection of minute cracks or minor inclusions.

4.2.5. AUTOMATIC APPARATUS FOR ONE-SHOT TESTING

Special purpose equipment for checking automatically large numbers of identical components of simple form can be designed in such a way that a current flow test and a coil test are applied at the same time, thus enabling both longitudinal and transverse defects to be found.

The two fields, if powered by a.c., tend to counteract each other but, *provided the resistance of the coil circuit is correctly adjusted*, the two fields are out-of-phase thus allowing detection of defects in both directions alternately. Defects, when delineated by ink particles, will still be revealed in spite of the demagnetizing effect of the other field after each cycle. However, detection with a.c. tends to be less sensitive than applying the separate tests and rectified a.c. excitation for this dual purpose is to be preferred.

A threading bar and coil have also been used in this manner and it would appear feasible for special applications to use other combinations of the basic testing methods if needed. For example, threading coil and induced current techniques could be used on rings. Apparatus of this type normally incorporates a conveyer system for progressing the parts.

4.3. FIXED INSTALLATIONS

4.3.1. POWER SOURCES

Since weight is no problem, fixed equipment can be designed with heavy-duty transformers having improved time on/time off ratios of about 1:3. Power sources include (*a*) d.c. batteries with heavy-duty chargers, (*b*) transformers, (*c*) transformers with full or half-wave inverters to give pulsated d.c. and (*d*) magnetic flux circuits powered by rectified a.c.

In selecting a power source, the following points must be considered:

(*a*) Pure d.c. has the disadvantages in that battery maintenance and battery replacement are expensive, but it is useful where the mains supply is non-standard because only the charger needs changing. An advantage is that a d.c. battery source gives deeper penetration than pulsating current and will find defects at a greater depth. However, the surface indication of deep-seated defects becomes hazy and, in general, defects below 0·100 in deep (2·5 mm) will be difficult to discover. As a comparison, the practical limit for a.c. would be about 0·070 in deep (180 mm) for the same sized defect.

(b) a.c. transformer machines are the cheapest and are usually chosen. Commercial types of up to at least 23,000A output are available.

(c) Machines using a.c. rectified to d.c. are expensive if a large output is needed. Their virtues are that they penetrate slightly deeper than pure a.c., give fewer false indications and always leave the component with full residual magnetism. This last feature is of value for production runs with similar parts, because the inking and viewing operations can be made after excitation.

(d) Magnetic flow machines are generally combined with current flow machines to give a method of producing a transverse flux in the component. They can vary widely in output due to accidental leakage in the yoke path and nowadays are tending to be replaced by built-in coil units which will give the same type of flux flow. They also have an advantage of avoiding any possibility of burning and are more suitable to provide adequate test levels for very massive components.

4.3.2. DESIRABLE MECHANICAL REQUIREMENTS

The desirable mechanical requirements are:

(a) The pole pieces should be faced with resilient copper gauze or some other provision should be made to prevent sparking during current flow testing. Rotation through 360° to facilitate viewing is a great advantage.

(b) The pole pieces should be of adequate size and be kept well away from the working surface so as to allow adequate clearance for the largest component.

(c) The pole gap must be able to accommodate the longest component. There should preferably be a motor or hydraulic drive, to adjust the gap, plus an adjustable pressure cut-off to ensure adequate squeezing of a large component without squashing a fragile one.

(d) An ink supply tank with mixer pump and hose feed should be provided. Ink tends to settle and stalling of the mixer after standing can be a problem.

(e) Provision should be made for easy attachment of coils, threading bars and wandercables. Many testing machines do have separately wired coils which should be adjustable sideways to improve efficiency in the testing of small components, since the coil centre has the weakest field.

(f) There should be no magnetizable steel in the ink system or in any position where jamming can be caused by adherence of the ink powder. It will be appreciated that any steel within the working area will be rendered magnetic.

4.3.3. DESIRABLE ELECTRICAL REQUIREMENTS

The desirable electrical requirements include the following:

(a) An ammeter of adequate size. For large equipment a high and low scale is often incorporated. The addition of an extra scale indicating peak values would be of value because most ammeters give r.m.s. readings.

(b) A multi-stud or continuous adjuster to allow current variation in steps of not more than 50 per cent rise or 500A (whichever is the least) from minimum to maximum.

(c) A foot or body operated switch. The operator may need both hands for manipulating the component. The switch should be adjustable in dwell of up to about 3 seconds and should operate automatically on the peak of the current cycle.

(d) A warning light to show that the ink pump is running.

(e) A charger, if fitted for d.c. battery machines; this should have a separate ammeter and light. The capacity, which may have to be as high as 30 amperes, must be adequate to maintain the batteries for maximum usage.

4.3.4. TYPICAL TESTING MACHINES

(a) Radalloyd Pin Machine (see Figure 4.1)

This is a typical small bench machine giving 1,000A a.c. with infinitely variable control. It is very suitable for testing pins, bolts and the like up to about 2 in (50 mm) diameter, provided that it is appreciated that transverse defects will not be revealed. In practice, bolt defects are almost invariably longitudinal.

(b) Radalloyd Model 32

This is a large 6 ft (18 m) machine with a peak current of 8,760A. A large coil is also available to give a magnetizing force of 1,800 oersteds (144 kA/m). The machine will test very large components (see *Figure 4.2*).

4.3. FIXED INSTALLATIONS

(c) Radalloyd Induced Current Machine

This is specifically used for testing rings of up to 4 ft (1·2 m) diameter and is quoted to give a peak current of 4,950A, which should suffice for rings up to 22 in (0·56 m) sectional perimeter to the suggested standard. It works on the principle of having a

Figure 4.1. Small hand machine for pins, bolts, etc. Alternating current flow infinitely variable up to 1000A (*By courtesy of Messrs. Radalloyd*)

detachable yoke so that the ring forms the secondary coil of the transformer. Defects found will be basically circumferential (see *Figure 4.3*).

(d) Fel-Electric Disc Tester

This is a very unusual machine using magnetic flux only. It was devised specifically for testing turbine engine discs but it is also

99

Figure 4.2. Model 32. 6 ft combined machine using coil, alternating current
flow and magnetic flow methods (*By courtesy of Messrs. Radalloyd*)

capable of testing rings having a massive section. The working area
is shown in *Figure 4.4.* The machine has four steel poles adjustable
in gap to accommodate discs of varying sizes. It has windings on
each pole to produce, at will, North pole or South pole magnetic
flux of sufficient strength to saturate a 36 in (0·90 m) disc 2 in (50
mm) thick. The shaped pole pieces are necessary to ensure intimate
contact with the components tested. By varying the polarity of the
four poles, the flux direction can be changed and defects in all
directions are detectable.

(e) Fel-Electric Model U2/ET

This is a popular machine which gives normally a peak of about
2,000A, although a greater output can be specified on ordering.
It comprises both current flow and magnetic flow circuits and has
provision for threading bars and adjusting the degree of squeeze
on the poles. An ink pump is fitted with a spray nozzle. The pole
gap is normally 3 ft (0·90 m). It will test all shapes of components
for defects in all directions.

(f) Fel-Electric Model LB 184/5

This is a powerful machine available with up to 5 ft (1·5 m)
pole gap and incorporates both Magnetic Flux and Current Flow
circuits. If desired, a peak current of 10,000A plus can be specified.

It is an excellent machine with a good ink pump and nozzle system plus powered heads and is very suitable for production run work, especially if the specialized coils which can be supplied are used (see *Figure 4.5*).

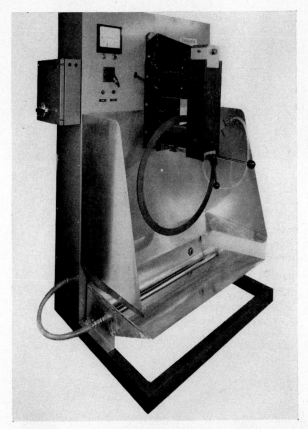

Figure 4.3. Ring tester using the induced current method
(*By courtesy of Messrs. Radalloyd*)

(g) *Fel-Electric Unimatic 167*

This is an example of a semi-automatic machine suitable for small work and testing rates of 720/h are possible (see *Figure 4.6*). The manufacturers will build automatic equipment, as desired, to suit any range of components.

Figure 4.4. Machine for testing discs or massive rings using the magnetic flux method (*By courtesy of Messrs. Rolls-Royce, Small Engine Division*)

Figure 4.5. Model LB184/5. Combined machine using alternating current flow and magnetic flow methods (*By courtesy of Messrs. Fel-Electric Ltd.*)

Figure 4.6. Uniamatic 167 (semi-automatic) (*By courtesy of Messrs. Fel-Electric Ltd.*)

(h) ESAB Model 167/1 or 2 Magnetox (see *Figure 4.7*)

This model is designed for production runs and has a pole gap of 5 ft 6 in (1·67 m) with either one or two 3,000A a.c. transformers, giving a peak of 8,640A for the double unit. A magnetic flux circuit of 107,000 oersteds (8,560 kA/m) is available. Refinements include rotating pole pieces and powered pole movement.

(i) ESAB Model 15-500, Aerotox (see *Figure 4.8*)

This equipment has a 7 ft (2·1 m) working gap with two transformers plus half wave rectifiers to give a d.c. of 8,640A and a magnetic flux circuit of 107,000 oersteds (8,560 kA/m). The pole pieces are extra large and an added refinement is a separately wired coil.

Figure 4.7. Magnetox Model 167/2 (*By courtesy of ESAB Ltd.*)

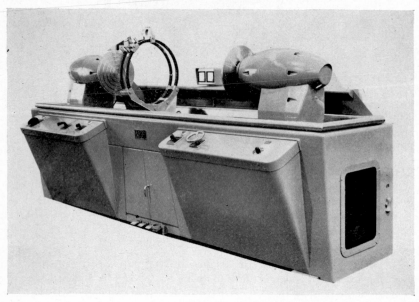

Figure 4.8. Aerotox Model 15-500 (*By courtesy of ESAB Ltd.*)

(j) *Magnaflux DRC543*

The manufacturers of this equipment can supply machines of all types of U.S.A. design. It is a 3,000A a.c. model with a built-in coil. Special features include adjustable squeeze on the pole pieces, a body or hand operated current switch and an automatic demagnetizing device which operates by progressively reducing the current in the coil.

(k) *Magnaflux 5,000A AC/DC Horizontal*

This is an example of a larger machine designed for testing heavy crankshafts. Extra features include a built-on black cloth hood to facilitate inspection using ultra-violet light and fluorescent ink.

(l) *Bar Machine*

Figure 4.9 shows a bar machine made by Metropolitan-Vickers (now part of the A.E.I. group). It is a most useful machine for testing bar stock up to 13 ft (4 m) long, although it will only show longitudinal defects. On raising the tank, which contains red ink, over the bar, the current is automatically adjusted and switched on.

Figure 4.9. Bar Tester for alternating current flow method only (*By courtesy of G.E.C.-A.E.I.* (*Electronics*) *Ltd.*

(m) *Metroflux Model S.C.—ex A.E.I.*

This is a large powerful machine with a 5,000A a.c. supply and magnetic flux circuit both controlled by variacs. An ink pump system is supplied and it is possible by using a d.c. reversing switch to demagnetize with the magnetic flow circuit (*see Figure 4.10*).

Figure 4.10. Metroflux Model S.C. for alternating current flow and magnetic flow methods (*By courtesy of G.E.C.-A.E.I. (Electronics) Ltd.*)

4.4. PORTABLE EQUIPMENT

(a) *Magnets*

Magnets are simple and cheap devices. Their use is restricted in general to testing specific suspect areas, such as rivet holes in a large tank or known trouble spots in a power station rotor. Suitable magnets include the Alcomax type in horse-shoe form as sold for radar applications. A typical example is the U. 175S giving 500 oersteds (40 kA/m) on a 2 in (50 mm) gap, as supplied by Messrs. Turton Bros. & Matthews. Aluminium shoes can be fitted to reduce flux when required. A model specifically designed with adjustable ends is the S.B. 127/B obtainable from Messrs. Fel-Electric. A sophisticated type with adjustable hinged ends can be supplied by Radalloyd

Ltd. Model 23A which has the addition of a mains operated winding to increase the field up to 600 oersteds (48 kA/m).

(*b*) *Battery Operated Testers*

A large car or aircraft battery can provide in excess of 500A, particularly if modified to give 2 V only by connecting the cells in parallel. A wandercable can be used as a feed to Gee-clamps on the work, as a coil or threading coil and as a temporary demagnetizer. Provided that the limitations of the power output are realized, adequate tests can be made on small parts, even *in situ*, in an aircraft or a road vehicle.

(*c*) *Transformers—A.C. Mains*

To transport A.C. mains transformers necessitates that the weight must be kept low. In their design, the ratio of time on/time off may be as low as 1:10, to reduce the weight of copper needed.

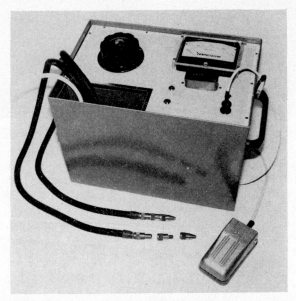

Figure 4.11. Compactaflux 2000-VA (*By courtesy of Stevic Engineering Ltd.*)

This is no snag, provided that the operator uses his discretion, does not leave the current on for more than about 3 seconds and allows time to cool between shots. Wandercables, etc. are needed as with a battery.

107

Commercial equipment, such as the Compactaflux equipment from Stevic Engineering who feature models giving 850A, 1,500A and 2,000A a.c. output, is readily available (see *Figure 4.11*). A mobile unit trolley mounted to give a fully variable output from the same source is shown in *Figure 4.12*.

Figure 4.12. Compactaflux 2000 ADVM (*By courtesy of Stevic Engineering Ltd.*)

Messrs. Fel-Electric also make robust portable equipment such as Model L.B.50 (1,000A a.c.) and the wheeled model, L.B. 124/3T which can provide a massive output of full adjustable 3,000A a.c. Messrs. Inspection Equipment make similar equipment and are marketing a plug in a.c. half-wave d.c. converter which can be used with their a.c. models.

4.5. AUXILIARY EQUIPMENT

In addition to the flaw detector and accompanying jigs, the equipment used in a factory will normally comprise the following:

4.5.1. DEMAGNETIZERS

(1) The usual form of demagnetizer is a rectangular coil powered by a.c. mains with sufficient windings to develop a magnetizing field higher than that left in any component by the testing machine.

This should have an opening sufficient to accommodate the largest component but, for economy and efficiency, it should not be over-large. For very large components, a wooden track and low platform can be fitted to facilitate handling.

The demagnetizer should ideally be positioned with the bore facing North and South, but it should always be placed to allow for the handling of the longest component, bearing in mind that this component should be brought at least 3 ft (1m) clear of the coil during the demagnetizing operation.

(2) It is possible to use the machine coils in lieu of a demagnetizer but this tends to be awkward and time consuming. Some machines do have a built-in automatic current reducer to facilitate demagnetizing.

(3) Wiper demagnetizers, which consist of a small coil powered through a cable, are available. They operate by stroking the component all over and then slowly withdrawing clear of the latter. This method is not really suitable for use for routine testing in a works but has advantages where massive structures are to be tested. One of the best known types is the Magnaflux Y-5 220 Yoke Kit which can also successfully serve as an electromagnet tester.

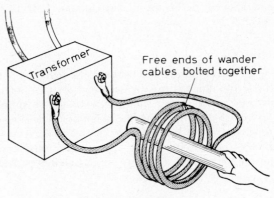

Figure 4.13. Emergency demagnetizer (*By courtesy of Messrs. Rolls-Royce Ltd., Small Engine Division*)

(4) For field work, it is possible to rig up a demagnetizer by using a transformer and wandercables as shown in *Figure 4.13.* In using this type of demagnetizer, it is important to rotate the component end to end several times during withdrawal to change the polarity if a battery is used as a source of power.

4.5.2. Demagnetization Check Equipment

To ensure that demagnetization is correctly performed, a compass test is specified for aerospace parts (see AvP. 970, Ch. 405 & 717 and A.I.D. leaflet AvP. 84, D.005). This test uses a standard P.4 compass (A.M. Ref. 6A/0745) and requires that when the component is rotated at an agreed distance from the compass, a maximum of 1° deflection is obtained. Since the distance is normally over 2 ft (600 mm) this is a requirement which is easily satisfied and it is feasible to demagnetize successfully; hence the compass can be placed 3 in (75 mm) from the component. For some applications such as electron-beam welding, this high standard is necessary to avoid deflection of the beam during welding.

A more convenient device is the Magnetic Field Indicator marketed by Magnaflux. It is quite cheap and of pocket watch size, and it will give one unit deflection equivalent approximately to a deflection of 1° on a compass at 3 in (75 mm). In addition, it is possible with this device to find on a complex shaped component local poles resulting from the configuration or local work, such as the vibro-marking of part numbers.

An *ad hoc* device, which is surprisingly effective, is a chain of eight interlooped wire paper clips. This is held vertically and passed round the component perimeter in close contact. It will adhere at a very low level of residual magnetism.

4.5.3. Lifting Gear

Heavy work can be suspended between the machine poles and a simple 4 to 1 rope hoist is usually sufficient for this purpose. For convenience, the work is best slung on to a jib or simple tripod so as not to foul the lighting over the machine. The slings are best made of rope covered with plastic sleeving because metal slings can be a nuisance due to 'magnetic writing' and the possibility of short-circuiting the current, with dire effects, if left in position during testing.

4.5.4. Viewing Aids

To facilitate viewing, the following items are useful:

(*a*) A small mirror, such as that fitted to ladies' handbags for viewing bores and the reverse side of a component.

(*b*) A $\times 2\frac{1}{2}$ to $\times 4$ hand magnifying glass, preferably illuminated. Messrs. P. W. Allen market an excellent range of this type of instrument.

4.5. AUXILIARY EQUIPMENT

(c) An Intrascope (or borescope, see Section 5.2.2.) which may be needed for viewing narrow bores.

4.5.5. DEGREASER

Prior to testing, the part should be cleaned. A standard trichlorethylene degreaser is normally used for this purpose in a works. In the field, *ad hoc* methods with rag and solvent must perforce be used. The use of trichloroethylene is not satisfactory because the fumes can be toxic, especially if the operator is smoking. White spirit, paraffin or methylated spirits are safer.

4.5.6. WASH TANK

Where many parts are tested, the removal of residual ink presents a problem. A paraffin wash tank, preferably with a hand pressure spray fed by a pump, is the usual answer. This should be supplemented with a long-handled brush to enable residual ink powder to be removed from crevices.

4.5.7. INK LADLE, ETC.

If an ink-feed hose and nozzle are not fitted, a standard stainless steel soup ladle will be needed to baste the component. Magnetic steel ladles should not be used. Alternatively, a polythene squeeze bottle or an oil can of the pump type, which should not be of magnetic steel, would be useful for site work. For dry powder work, powder applicator bulbs are commercially available.

4.5.8. VIEWING GOGGLES

When using black light for fluorescent ink viewing, discomfort can be caused by ultra-violet light reflected into the operators' eyes. The eyeballs will fluoresce and headaches can result. A simple cure is to use either welders' goggles or a motor-cycle type eyeshield fitted with a minus blue filter, which will stop the ultra-violet but pass light of wavelengths between 550 nm and 600 nm (5500 and 6000 Å). Suitable types are the Ilford 108, Kodak Wratten 4A or I.C.I. Perspex VA can also be obtained with a suitable transmission band. The viewing contrast is actually improved by using such a filter.

4.6. SITE AND LIGHTING REQUIREMENTS

For the operator to concentrate on his work, the site should be reasonably quiet, well heated, and away from the direct traffic flow in a busy shop. If possible, it should be situated near the machine shop and also near the viewing room to simplify transport.

Main lighting *must* be adequate. Natural light is not essential provided that a strong even light is available over the full length of the machine. This requirement is most easily fulfilled by having two 5 ft (1·5 m) fluorescent tubes coupled out-of-phase and mounted about 6 ft (2 m) from the ground directly over the axis of the magnetic poles. The reflector should be a single one and be wide enough for shadows and highlights on the test components to be equally avoided. The level of lighting at all points in the viewing area should not be less than 500 lux.

A supplementary hand lamp is often needed to view inside bores. The narrow diameter 1 ft (0·3 m) fluorescent hand lamps, as used by garages, are probably most suitable.

If fluorescent magnetic ink is to be used, the above requirements must be supplemented by the use of a filtered ultra-violet or ' black' light to give a minimum intensity of about 100 lux. There is considerable discussion taking place at present at B.S.I. to define an agreed minimum standard and to recommend an official method of measurement as discussed below. The predominant wavelength required is 365 nm (3650 Å). The ideal arrangement is to use both a hand lamp and a 5 ft (1·5 m) special ultra-violet fluorescent tube (or tubes) placed above the machine.

Figure 4.14. Model 16, 'black light' hand lamp (*By courtesy of Engelhard Hanovia Lamps Ltd.*)

4.5. AUXILIARY EQUIPMENT

An auxiliary ultra-violet hand lamp such as the Model 16 (Hanovia) is necessary for bore viewing and to provide intense light for local viewing. For occasional work only, this could be adequate without using the 5 ft tube mentioned above (see *Figure 4.14*). Messrs. P. W. Allen can supply the alternative types shown in *Figures 4.15* and *4.16*.

Figure 4.15. Model A.408, 'black light' lamp (*By courtesy of P. W. Allen & Co.*)

Figure 4.16. Model A.409, 'black light' lamp, designed for viewing small components on the bench (*By courtesy of P. W. Allen & Co.*)

A necessary corollary to the use of 'black light' is a booth surrounding the machine so that ordinary white light is almost totally obscured (less than 5 lux at the viewing position). A hardboard roof and black curtains hung from a four sided rail are commonly used. To assist in ventilation and to prevent claustrophobia, it is best to hang the curtains about 1 ft (0·3 m) short of the floor. The white light thus allowed to enter the booth will be confined to the floor area and will not spoil viewing conditions. An improvement on using all-round curtains is the use of aluminium panels for three sides. These panels have the advantages of keeping cleaner than curtains and also of reflecting the available ultra-violet light back to the test area to provide stronger and more even lighting.

4.7. VERIFICATION OF APPARATUS AND TESTING

4.7.1. CURRENT FLOW MACHINES

Current flow machines must be fitted with an ammeter which should read within 3 per cent of the true value at one quarter, half, three-quarters and full scale. This condition is easily attained by ammeters of the usual commercial standard. It is suggested that the accuracy should be verified at least every 3 months by using a check meter having a maximum error of 2 per cent. It is particularly important to verify exactly what the ammeter is reading, because it is the 'peak' current which is important.

An ammeter fitted to a d.c. (battery) source will read the average current which is also the peak current. An ammeter fitted to an a.c. or d.c. (full wave) source will read the average r.m.s. (root mean square) value and the peak current will be $\sqrt{2}$ times this value. An ammeter fitted to a d.c. (half-wave) source is normally adjusted to read r.m.s. current and the peak current is again $\sqrt{2}$ times this value. The average for half-wave d.c. is actually half-r.m.s. current.

The maximum output possible from a current flow machine is usually quoted as the 'shorting-value', which is seldom reached in practice, because the resistance of a test component reduces the output. The output can also be reduced by the choke effect caused by the configuration of a complex shaped component. As an example, on one specific machine with a shorting value of 15,000A, the use of two 12 ft (about 4 m) wandercables reduced the maximum output at the cable ends to 4,100A. Thus the power requirements for a given component may necessitate the use of a source having an apparently excessive output. A qualified electrician should verify the electrical safety of the machine every three months.

To determine the 'shorting-value' maximum output for comparing machines, a standard bar should be used (see *Figure 4.17*).

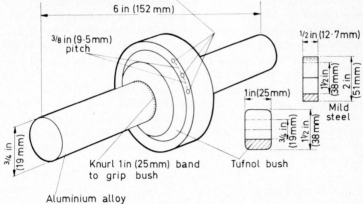

Figure 4.17. Standard test piece for the current flow method (*By courtesy of Messrs. Rolls-Royce Ltd., Small Engine Division*)

4.7.2. MAGNETIC FLOW MACHINES

For testing new magnetic flow machines or for comparing these machines, an oersted meter, which normally works for a 3 in (75 mm) gap between the poles, can be used. By adjusting the energizing current, the flux output of a specific machine can be correlated to the ammeter reading. This is considered important since equipment of the same type has been found to vary considerably due to minute differences in machining allowances in various parts of the yoke circuit. A repeat check at regular intervals is also advisable because wear can affect the output of flux at the poles.

The recommended method of using flux machines is to adjust the flux level for the component to the furring (saturation) level and then to reduce it to about 40 per cent of this value for testing. Thus the output of the flux circuit is checked automatically, but there is a requirement that the flux level/ammeter reading ratio throughout the range should be known.

4.7.3. MAGNETIC INK

Assuming that ink complying with B.S. 4069 is used, there are two possible faults for which to check. These are:

(*a*) Loss of powder due to settling in the machine or pump crevices.

(*b*) Contamination due to oil or water pick-up causing coagulation and loss of sensivity.

A simple test is to take a sample from the ink, as used, in a 100 ml Crowsfoot receiver. This is a standard graduated tapered-base glass cylinder. After thirty minutes to allow settling, the level of powder can be checked and this should be of the same order as that for new ink (usually $1\frac{1}{2}$ per cent or upwards). The colour of the liquid will be brown if oil contamination is present. Separation globules will be visible if water is present.

A much more sensitive test is to use a prepared test-piece such as that shown in *Figure 4.17*. This will readily show a loss of sensitivity of the ink. Ink in good condition should show all three hole indications when tested at 1,000A (peak) and if a test at 1,500A does not show all three indications, the ink should be changed immediately.

4.7.4. 'BLACK LIGHT'

The wavelengths of ultra-violet light, as used for fluorescent viewing, are so short that minute amounts of dust or reflector tarnishing can reduce the output drastically. Regular maintenance at about three-monthly intervals is needed. This includes cleaning the reflector, wiping the tube surface and, in particular, removing adherent film from the reflector. A suitable device to check efficiency is, at the time of writing, being discussed at B.S.I. and recommendations should be available shortly. The device consists of a standard x-ray fluorescent screen, which responds strongly to ultra-violet light, and a photocell to measure the response at the viewing position.

Pending the standardization of this device, the lamp strength can be verified as follows:

(*a*) By viewing a known test-piece with the suspect lamp and with a new lamp of known characteristics.

(*b*) By using a luminous dial wrist watch as a standard. This can be placed at the viewing position and the furthest distance at which it can be seen distinctly will give a measure of any major fall-off in intensity.

(*c*) By using a filter paper soaked in fluorescent penetrant solution at the viewing position and measuring the converted light output with a Weston photographic meter. A minus blue filter (see Section 4.5.8) should be placed over the screen. Readings will be of the order of 12 foot-candles (120 lux) with a poor light and as high as 50 foot-candles (500 lux) with a good light.

4.7.5. CHECKING EFFICIENCY OF A TEST ON A GIVEN COMPONENT

With complex shapes, it is sometimes difficult to be sure that the flux flow is in the desired direction and is of sufficient strength in certain areas. The following procedures are applied, where relevant:

(a) The power input into the part can be increased until saturation (furring) occurs. The flux-flow direction will be visible and the desired level of saturation can be worked out from the power required for saturation.

(b) With magnetic flow or coil tests, North and South residual poles will be present in the component after the examination. The presence of these poles can be verified and their strengths determined with either a magnetic field indicator or compass. A reading of over 5 units on a magnetic field indicator is considered acceptable.

(c) A component can be drilled and the hole filled with a plug of the same material. A deliberate clearance of 0·002 in (0·05 mm) will give a detectable defect in the form of a circle. The segments of the circle which are indicated will show the flux direction and the clarity of the indications will improve with the efficiency.

(d) Portable defects can be attached to components with tape or pressed on to the surface at key positions. These can be obtained

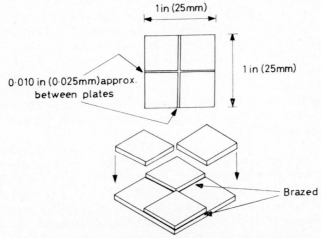

Figure 4.18. Test block suitable for use with current flow, magnetic flow, or coil methods (*By courtesy of Messrs. Rolls-Royce Ltd., Small Engine Division*)

117

commercially, e.g. the Exorel MCD Gauge as supplied by Inspection Equipment Ltd. which consists of a ring fitted with a fork handle. This ring contains four soft iron segments of a circle. The clarity of the division lines indicate the flux direction and intensity. A drawing of a suitable portable indicator working on this principle is shown in *Figure 4.18*. An even simpler and cheap device which can be very

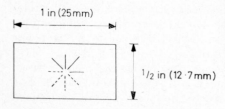

Figure 4.19. Spring steel portable flaw test-piece showing lines produced by an electic arc engraver (bold lines for heavy setting and broken lines for light setting) (*By courtesy of Messrs. Rolls-Royce Ltd., Small Engine Division*)

efficient is illustrated in *Figure 4.19*. A number of these can be made quickly from feeler gauge strip and taped in key positions on the test component. Both these test-pieces should be used face downwards.

(*e*) Oersted meters are available commercially and, although expensive, can give a complete picture of flux direction and strength.

4.8. THE OPERATOR

The requirements for a good operator include the following:

(*a*) *Good eyesight.* This is essential if he is to see the small indications likely to be found. There are at present no U.K. standards for inspectors' eyesights although the requirements are at present under discussion for aerospace standards. It is good practice for the operator to wear eye glasses, if he needs them, and to be capable of reading small print under the lighting conditions which pertain to the machine he uses. Colour blindness is probably of no hindrance, unless the operator is using fluorescent ink and is suffering from the rare 'green blind' form.

There is a U.S.A. Standard MIL-STD 401 which calls for distant vision to be 20/30 in at least one eye, either uncorrected or corrected. It also calls for near vision to be equal to normal vision in at least one eye, either corrected or uncorrected.

(*b*) *Conscientiousness.* None of the operator's work is normally checked to verify if all significant defects have been found. In some large engineering works, the bonus system applied can tempt an operator to leave out some part of the full procedure. For this reason, any payment for increased output is probably not advisable.

118

(c) *Good training.* The principles of magnetic particle testing must be understood to ensure an adequate inspection and any new operator should be instructed in the basic requirements before being left to test and accept work. A specified technique for each component, however simple, is always advisable.

(d) *Cleanliness.* The paraffin base which is usual for magnetic ink does not have a direct health hazard. However, continuous immersion of the hands in paraffin oil can have a softening effect on the skin and thus encourage dermatitis to be picked up from other sources. Mild cases of dermatitis have occurred where operators have not taken reasonable precautions and have been careless about washing.

4.9. JIGS AND FIXTURES

It should be remembered that jigs and fixtures intended to carry current should not be made of magnetic material but should be made of a material having a copper or aluminium base of adequate section to carry the heavy current used. Jigs which carry magnetic flux, however, must be made of massive magnetic material (usually mild steel) and should not be made of a material having a copper or aluminium base.

4.9.1. THREADING BARS

For threading bar tests, aluminium rods or plugged tubing are best for lightness. There is a slight risk of local burning in bores if internally threaded parts, springs or other irregularly contacting parts are tested. It is advisable, where such parts are tested, to insulate the outside diameter of the threading bar with a plastic sleeve.

A useful range of dimension includes the following:

Plain bar (a) $\frac{1}{2}$ in (12 mm) × 18 in (0·46 m)
         (b) $\frac{3}{4}$ in (19 mm) × 18 in (0·46 m)
         (c) $\frac{5}{16}$ in (8 mm) × 18 in (0·46 m)
         (d) 1 in (25 mm) × 30 in (0·76 m)
         (e) $1\frac{1}{4}$ in (32 mm) × 30 in (0·76 m)

(Above this size, solid bar is too heavy and a plugged tube is needed, see *Figure 4.20*).

         (f) $2\frac{1}{2}$ in (64 mm) × 4 ft (1·2 m)
         (g) 3 in (76 mm) × $4\frac{1}{2}$ ft (1·4 m).

Larger sizes are generally unnecessary.

119

Flexible flat threading bars such as those illustrated in *Figure 4.21* are useful for passing through shallow apertures. For testing very small work, such as nuts, the small diameters of threading bar necessary will cause sagging if the bar is placed between the machine

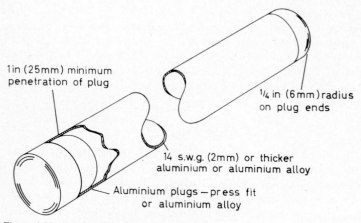

1 in (25mm) minimum penetration of plug

¼ in (6mm) radius on plug ends

14 s.w.g. (2mm) or thicker aluminium or aluminium alloy

Aluminium plugs — press fit or aluminium alloy

Figure 4.20. Lightweight heavy duty threading bars (*By courtesy of Messrs. Rolls-Royce Ltd., Small Engine Division*)

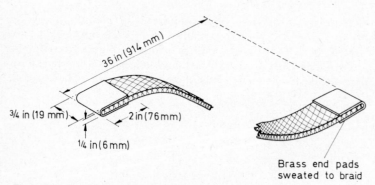

36 in (914 mm)

¾ in (19 mm)

2 in (76mm)

¼ in (6mm)

Brass end pads sweated to braid

Figure 4.21. Flexible threading strap (*By courtesy of Messrs. Rolls-Royce Ltd., Small Engine Division*)

poles. *Figure 4.22* illustrates a simple jig to avoid this. The jig will accommodate several hundred small nuts in one load. It should be remembered that since four threading bars are used simultaneously, the current must be four times higher.

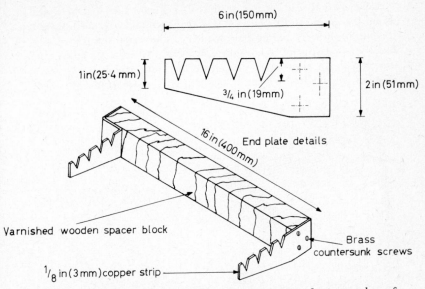

6 in (150 mm)

1 in (25·4 mm)

³/₄ in (19mm)

2 in (51mm)

End plate details

16 in (400 mm)

Varnished wooden spacer block

Brass
countersunk screws

¹/₈ in (3 mm) copper strip

Figure 4.22. Multiple bar tester. Additional requirements are four copper bars of ¼ in (6 mm) or alternative diameter to suit the work to be tested. (*By courtesy of Messrs. Rolls-Royce Ltd., Small Engine Division*)

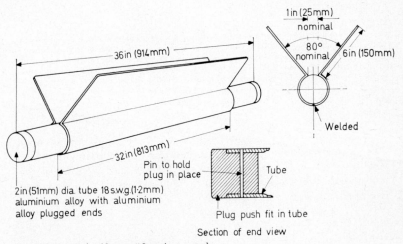

1 in (25mm)
nominal

80°
nominal

6 in (150mm)

36 in (914mm)

Welded

32 in (813mm)

Pin to hold
plug in place

Tube

2 in (51mm) dia. tube 18 s.w.g. (1·2mm)
aluminium alloy with aluminium
alloy plugged ends

Plug push fit in tube

Section of end view

Jacket (mild steel x 18 s.w.g. (1·2mm) or near)
cadmium plated push fit on tube

Figure 4.23. Jacketed threading bar (*By courtesy of Messrs Rolls-Royce Ltd., Small Engine Division*)

121

A jacketed threading bar as shown in *Figure 4.23* will be found useful for testing short items such as studs which are placed to span the gap in the vee. A number of studs can be tested simultaneously, but the current applied must be consistent with the threading bar diameter. An alternative procedure for bolts or studs is to test a number in parallel by using a jig such as that shown in *Figure 4.24*.

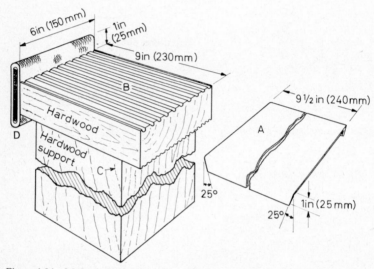

Figure 4.24. Multiple bolt tester. A—Sliding Contact plate in thick copper sheet with sliding fit on wood block. B—Top surface of wood block with $\frac{1}{4}$ in (6 mm) grooves with $\frac{3}{8}$ in (10 mm) lands. C—Bottom surface (reversible) with $\frac{1}{8}$ in (3 mm) grooves with $\frac{1}{4}$ in (6 mm) land. D—End pad is a 40 mesh 4 layer copper wire cloth with $\frac{1}{4}$ in (6 mm) oil resistant rubber centre, fixed with brass screws (*By courtesy of Messrs. Rolls-Royce Ltd., Small Engine Division*)

4.9.2. SUPPORT BLOCKS

Many items which require testing are heavy and vee-notched. The use of wooden support blocks tailored to suit main items are well justified, since they facilitate the loading and rotating of the component. Varnished wood is a suitable material and typical examples are shown in *Figure 4.25*.

4.9.3. COILS

Fixed coils for use on the machine can be either squeezed directly between the pole pieces or, if large, separately connected by heavy cables. The material used is almost invariably flat copper strip about 1 in (25 mm) wide by $\frac{1}{8}$ in (3 mm) thick and is usually

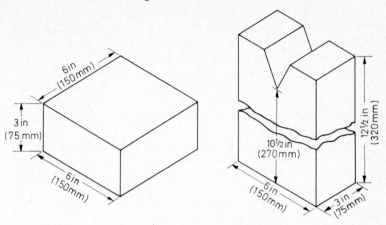

Figure 4.25. Support blocks (*By courtesy of Messrs. Rolls-Royce Ltd., Small Engine Division*)

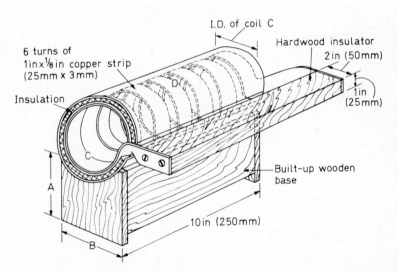

Figure 4.26. Small coils. A suitable design of 4 in (100 mm) and 8 in (200 mm) coils for direct connection by squeezing between machine poles. A = 6 in (150 mm) or 5 in (130 mm) respectively. B = 3¾ in (95 mm) or 7½ in (190 mm). C = Helix lined with insulating sheet to avoid short-circuiting. D = Steel concentrator in heavy mild steel sheet insulated from copper helix and must have a gap along the length (*By courtesy of Messrs. Rolls-Royce Ltd., Small Engine Division*)

123

formed into a helix of 4 or 5 turns. Fewer turns than this will give a lower field. More turns than this will also give a lower field, because the increased electrical resistance and reluctance (i.e. magnetic flow 'resistance') reduces the current in the coil.

The usual shape of coils is circular but there are definite technical advantages in flattening a coil if the shape of the component allows this. All coils should be insulated except at the contact ends otherwise the component could be burned. The use of a plastic coating is usual.

Although not essential, the provision of a mild steel wrapping around the coil, and insulated from it, has been found to improve efficiency by approximately 50 per cent by facilitating the passage of flux through the air return path. The jacket must have a slight gap in the circumference. *Figure 4.26.* shows a typical design of two small coils with wooden spacers to enable them to be squeezed between the magnetic poles. *Figure 4.27* shows a typical flat pancake coil fitted with aluminium trays which facilitate the testing of small bolts and the like. Testing a number of bolts simultaneously

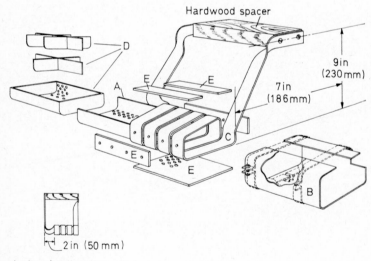

Front view in direction of A

Figure 4.27 Pancake coil. A = Insulated coil liner of fibre sheet. B = Heavy gauge mild steel intensifier with gap insulated by E from coil to cover length. C = Pancake coil from $1\frac{1}{4}$ in (32 mm) × $\frac{1}{8}$ in (3.2 mm) copper strip with inside dimensions $5\frac{1}{2}$ × $5\frac{1}{2}$ × 2 in (140 × 140 × 50 mm). D = Perforated aluminium work tray with 4 way and 6 way spacers 5 × 5 × $1\frac{1}{2}$ in (127 × 127 × 38 mm) (*By courtesy of Messrs. Rolls-Royce Ltd., Small Engine Division*)

improves the fill factor and efficiency of the test. Very large coils are usually free standing and connected to the power source by short heavy cables. *Figure 4.28* shows suitable examples, progressing in 4 in (100 mm) consecutive steps.

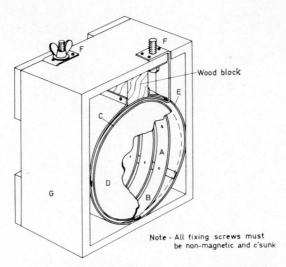

Figure 4.28. Large free-standing coil, suitable for wander-cable connection to machine poles. A = 4 full turns of twin $1\frac{3}{4} \times \frac{1}{8}$ in (45 × 3 mm) copper strip, $\frac{1}{2}$ in (13 mm) edge spacing. B = Insulated liner in coil. C = Insulation to prevent concentrator touching coil. D = Diameter as needed. Suitable steps are 12, 16, 20, 24 in (300, 400, 500, 600 mm). E = Heavy gauge mild steel concentrator with gap. F = Connection bolts to suit 2 off 30 in (750 mm) cables in 2200/.0076 in with BS.9 Type 1 13E cable ends. G = 1 in (25 mm) nominal stained and varnished wooden frame (*By courtesy of Messrs. Rolls-Royce Ltd., Small Engine Division*)

For testing components across the axis in a coil, it is essential to use mild steel extenders to improve the *L/D* ratio. *Figure 4.29* shows a typical example in which an aluminium tray facilitates the positioning of mild steel plate extenders. It is sometimes advantageous to make up a coil specifically to suit a specific component. A typical example of design used with wandercables, is shown in *Figure 4.30*. The coil was specifically made to fit over a shaft. Occasionally, a test component such as a ring needs an openable coil which can be placed round a section and powered through

125

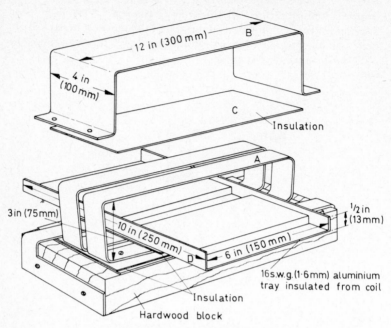

Figure 4.29. Coil with built-in steel extenders. A = Rectangular coil of two or more turns from 1 × ⅛ in (25 × 3 mm) copper strip with ¼ in (6 mm) gaps. B = Heavy mild steel concentrator screwed to base. C = Insulation to stop B touching A. D = 16 G (1·6 mm) aluminium tray insulated from coil with 2 off mild steel extenders 6 × 4 × ⅜ in (150 × 100 × 9 mm) to slide inside (*By courtesy of Messrs. Rolls-Royce Ltd., Small Engine Division*)

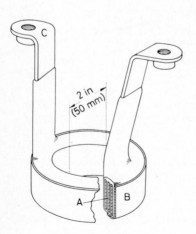

Figure 4.30. Typical coil to fit a specific shaft. A = 3 turn coil from 1 in × 16 S.W.G. (25 × 1·6 mm) copper strip insulated with glass fibre and epoxy resin. B = Mild steel concentrator 18 S.W.G. (1·2 mm). C = Suitably threaded connector (*By courtesy of Messrs. Rolls-Royce Ltd., Small Engine Division*)

wandercables. *Figure 4.31* shows a suitable one-turn design. Multi-turn versions, which are less demanding for current, are available commercially.

4.9.4. WANDERCABLES

For the testing of free-standing objects of large size, it is usual to connect heavy cables directly to a transformer or to machine poles. To avoid excessive current drop, these cables should be as massive and as short as feasible. Standard welding cable, such as B.S. 638 2200/·0076 T.R.S. or 610/·018, is suitable and lengths of up to 12 ft (3·7 m) long are used. *Figure 4.32* shows a simple connecting block for connecting them to machine poles. The cables can be used as threading coils and threading bars and can also carry current to the connecting clamps. If necessary, they can be formed into temporary coils or even used as demagnetizers (see *Figure 4.13*).

4.9.5. CONNECTOR CLAMPS

To carry the current from wandercables into test objects, the end connectors must be massive copper gauze faced clamps. Ordinary G-clamps, suitably gauze-lined, or more specialized clamps such as those shown in *Figures 4.33* and *4.34* can be used. A more sophisticated clamp, termed a magnetic leech, is very suitable for flat surfaces.

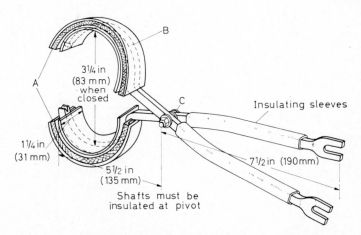

Figure 4.31. Openable coil for use with wandercables. A = Single turn coil with insulated joint and copper handles (insulated internally). B = Mild steel concentrator insulated from coil. C = Brass nut and bolt (*By courtesy of Messrs. Rolls-Royce Ltd., Small Engine Division*)

127

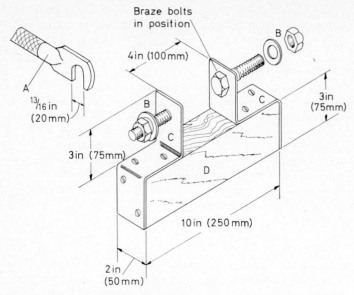

Figure 4.32. Connecting block for wandercables. A = Wandercable end. B = $\frac{3}{4}$ in Whit. brass bolts (19 mm) with nuts. C = Hard drawn copper strip 2 × $\frac{1}{8}$ in (50 × 3 mm) with brass fixing screws. D = Hardwood block, stained and varnished (*By courtesy of Messrs Rolls-Royce Ltd., Small Engine Division*)

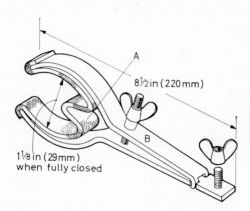

Figure 4.33. Screw clamp for use with wandercables (current flow method). A = Copper wire cloth, 50 mesh × 35 S.W.G. (0·2 mm). B = Steel or aluminium strip (*By courtesy of Messrs. Rolls-Royce, Ltd., Small Engine Division*)

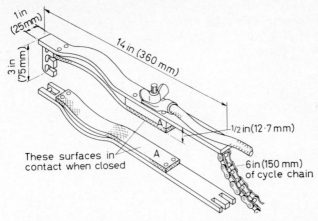

Figure 4.34. Adjustable clamp for use with wandercables (current flow method). A = Copper wire cloth, 50 mesh × 35 S.W.G. (0·2 mm) (*By courtesy of Messrs. Rolls-Royce Ltd., Small Engine Division*)

It uses a horseshoe magnet which carries an insulated brass and copper contact pad (see *Figure 4.35*).

4.9.6. COMPONENT FITTINGS

Components which have small cross-sections at the ends can provide problems when current flow tests are needed. The use of copper

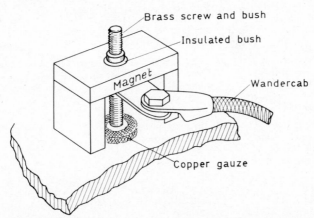

Figure 4.35. Magnetic leech for use with wandercable and current flow test (*By courtesy of Messrs. Rolls-Royce Ltd., Small Engine Division*)

gauze pads is one answer but the employment of specialized end-fittings to suit each component is preferable. *Figure 4.36* shows such a fitting which also allows rotation of the component during testing.

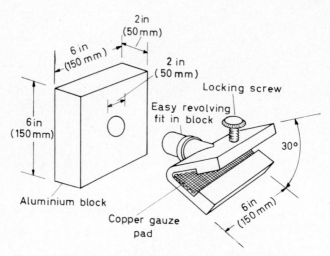

Figure 4.36. Typical end-fitting for taper-end component (*By courtesy of Messrs. Rolls-Royce Ltd., Small Engine Division*)

4.10. SUPPLIERS OF EQUIPMENT

U.S.A. AGENTS

P. W. Allen & Co. 253 Liverpool Road London, N.1 Tel.: 01-607 4665	Mechanical Technology Inc. 968 Albany Shaker Road Latham, N.Y. 12110 518785-0922	Inspection lamps, including 'black light' magnifiers
Engelhard Hanovia Lamps Bath Road, Slough Burnham, Bucks. Tel.: Burnham 4041	Engelhard Hanovia Inc. 100, Chestnut Street, Newark, N.J. Tel.: Bigelow 2-24-22	'Black light' lamps
ESAB Ltd. Gillingham, Kent Tel.: Medway 35261	L. E. Baxter Ltd. 6077 St. James Street West, Montreal 28, Quebec, Canada Tel: (514) 489-5951	Large and special purpose current flow and magnetic flow machines

4.10. SUPPLIERS OF EQUIPMENT (*cont.*)

UK Supplier	U.S.A. AGENTS (*cont.*)	Description
Fel-Electric Ltd. Leyburn Road, Sheffield, S8 OXA Tel: 0742 27257/8		Current flow and magnetic flow machines, magnetizers, magnetic ink, jigs and fixtures. Special machines and equipment to order
G.E.C.-A.E.I. (Electronics) Ltd., Scientific Apparatus Division, Barton Dock Road, Urmston, Manchester Tel: 061-865 4466		Current flow and magnetic flow machines. Magnets, demagnetizers, magnetic ink
Inspection Equipment Ltd. Oaklands Road, London, N.W.2 Tel: 01-452 8282		Current flow machines. Demagnetizers
Magnaflux Ltd. 702, Tudor Estate, Abbey Road, London, N.W.10 Tel: 01-965 5359	Magnaflux Testing Systems Corp., 7300, West Lawrence Ave., Chicago, Ill. 60656	Machines of all sizes and types. 'Black Light' lamps, magnetic powder, ink and accessories
Radalloyd Ltd. Glen Road, Oadby, Leicester Tel: Oadby 2531		Current flow machines, large and portable magnetizers. 'Black light' hand lamps
Stevic Engineering Woodlands Works, Water End Road, Potten, End, Berkhamsted, Herts. Tel: Berkhamsted 5646		Current flow machines, especially portable and transportable types.
Turton Bros. & Matthews Ltd., P.O. Box. 40, Rutland Road, Sheffield, S3 9 PL Tel: 0742 23156		Permanent magnets

5

OPTICAL METHODS OF TESTING

5.1. GENERAL CONSIDERATIONS

THE simplest non-destructive test of a component is to look at it either with the naked eye or with some aid to vision such as a magnifying glass or microscope. A simple visual test can reveal gross surface defects thus leading to an immediate rejection of the component and consequently the saving of much time and money which would otherwise be spent on more complicated means of testing.

However, it is often necessary to examine a surface for the presence of finer defects and, for this purpose, visual methods have been developed to a very high degree of precision. These methods include the use of magnetic particle and penetrant techniques, which are discussed at some length in other chapters. In many cases, especially for the examination of small components and of high-grade materials and also where it is undesirable to contaminate the surface, the uses of optical techniques are more suitable.

With the exception of *holography*, which is not yet in general use but is likely to have a promising future, only those methods which have some bearing on the routine testing of engineering materials are dealt with here. More sophisticated techniques, suitable for the research laboratory, are omitted but an excellent account of some of them has been provided by Heavens[1].

It is outside the scope of a work of this size to provide detailed descriptions of the various methods of testing but a sufficient account is given of the physical principles involved in order that the reader should have a clear idea of the techniques. Fuller details of the constructions of the various devices and their applications may be obtained by consulting McMaster[2] and the other references given in the text.

The principal optical methods of testing materials may be conveniently classified into two categories, namely:

(1) The use of instruments such as microscopes, telescopes and projectors, which apply the principles of geometrical optics, by which the rays of light are said to travel in straight lines;

(2) Techniques such as interference, photoelasticity, and holography, which involve the use of physical optics, in which the wave nature of light is concerned.

In common with other methods of inspecting surfaces, it is essential with optical techniques that the sample is properly prepared prior to testing, by the removal of dirt, grease, corrosion, and other surface impurities. It might be necessary for the surface to be etched and polished but it is always essential that it be properly illuminated to provide the most favourable conditions for examination.

5.2. APPLICATIONS OF GEOMETRICAL OPTICS TO THE EXAMINATION OF MATERIALS

Instruments used for the application of geometrical optics, including microscopes, projectors, and telescopes, are described in some detail by Martin[3] and by Habell and Cox[4]. In this section a brief description of their basic principles is given.

5.2.1. MICROSCOPES

Minute defects and details of fine structure in surfaces can be detected more easily with the aid of a microscope. The basic form of this instrument is the *simple microscope*, commonly called the magnifying glass. This is a single convex lens of short focal length placed close to the eye. The object under examination is situated just inside the focal point P (see *Figure 5.1*). An enlarged virtual image appears to be formed at a greater distance from the lens. The *magnifying power* (or *magnification*) M of the lens is defined as the ratio of the linear size of the image to that of the object and can be shown to be equal to the ratio of the distance, from the lens, of the image to that of the object. It can also be shown that

$$M = d/F + 1 \tag{5.1}$$

where d is the distance of the image from the lens and F the focal length. In practice d has a value of about 250 mm (10 in), the closest distance for distinct vision, at which the normal eye can see most clearly. *Figure 5.2* illustrates a modification for which the surface under test can be compared directly with a standard defect-free surface. A glass plate, silvered only over one half of its surface, is inclined at an angle of 45 degrees to the plane of the lens, thus enabling the adjacent images of the two surfaces to be viewed.

The practical upper limit of the magnifying power of a simple microscope is in the region of $10\times$, which corresponds to a focal length of 30 mm (about 1·2 in). An increase in magnification involves a decrease in focal length and, consequently, an increase in curvature of the lens surfaces. A lens having a large curvature tends to produce a distorted and blurred image. Furthermore, such a lens must of necessity be thick and the clearance between the lens surface

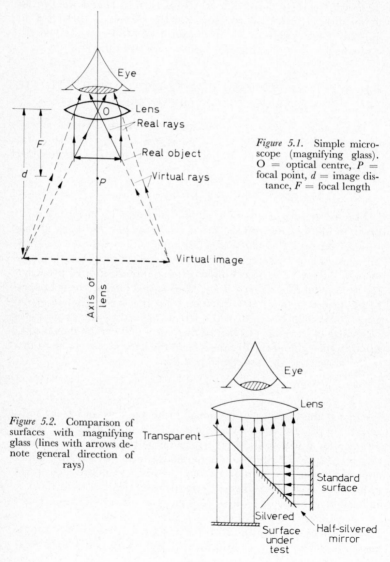

Figure 5.1. Simple microscope (magnifying glass). O = optical centre, P = focal point, d = image distance, F = focal length

Figure 5.2. Comparison of surfaces with magnifying glass (lines with arrows denote general direction of rays)

134

and the focal point F, which must contain the object, may be too small to be of any practical use for non-destructive testing applications.

Greater detail can be observed with the use of a *compound microscope*, such as the one illustrated in *Figure 5.3*, which may well have a magnifying power of several hundred times. The final image is formed by an arrangement of two lenses, or systems of lenses, called

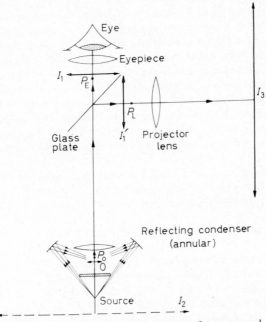

Figure 5.3. Simplified optical arrangement for compound microscope, with built-in projector, used for the examination of opaque surfaces (not to scale). O = object under examination, I_1 = real image of O formed by objective, I_2 = virtual image of I_1 formed by eyepiece, I_1' = real image of O formed by objective after reflection at glass plate, I_3 = real image of I_1' formed by projector lens on translucent screen or photographic plate, P_O, P_E, P_L = focal points of objective, eyepiece, and projector lens, respectively

the *objective* and *eyepiece*, respectively. The objective, which is placed near to the surface under examination, forms a magnified real image just inside the focal point of the eyepiece. The eyepiece is used to view this real image, in the same way as a magnifying glass, and the final image, which is enlarged still further, is seen at the closest

135

distance for distinct vision. Blurring and distortion are reduced by the use of systems of two or more lenses in each of the objective and eyepiece. The object is illuminated by light rays passing through a lens system called a *condenser*. *Figure 5.3* shows the principle of the metallurgical type of microscope as used for the examination of the surfaces of opaque objects. The figure also shows a modification which can provide projection of the image on to either a translucent screen or a photographic plate.

Generally speaking, the amount of detail of surface structure which can be seen is increased by amplifying the magnification with the use of lenses of higher magnifying power. Increasing the magnification in this way, however, has two major disadvantages. Firstly, there is a decrease in the size of the field of view, which means that the examination becomes a more tedious process. Secondly, there is the limitation due to what is called the *resolving power* of the instrument. Once the magnification has reached this limitation, any further increase in magnifying power results in the appearance of diffraction patterns and no further detail in the object can be observed.

The resolving power (R.P.) of the microscope is the smallest distance between two neighbouring points of detail on the object which can just be distinguished clearly. It can be shown that

$$\text{R.P.} = 0.5\lambda/n \sin \alpha = 0.5\lambda/\text{N.A.} \qquad (5.2)$$

where n is the refractive index of the medium in which the object is situated, α the half-angle subtended to the point, where the object intersects the axis of the objective, by the diameter of the aperture of the objective, and λ the wavelength of the light. N.A. $= n \sin \alpha$ is called the *numerical aperture*. It can be seen that the degree of resolution can be increased (by decreasing R.P.) if one employs short wavelengths and uses a large aperture objective. The resolution can be improved still further by increasing the refractive index of the space between the objective and the surface under examination (see Section 5.3.1) by what is known as *oil immersion*. The use of cedar-wood oil, which has a refractive index of 1.516 for sodium light, has the effect of stepping up the resolving power by about 50 per cent. In this way, a magnifying power of the order of 500× is sometimes possible without loss of resolution.

The most powerful type of microscope in existence, the *electron microscope*, does not employ visible light. Instead, it uses electromagnetic waves generated by the propagation of fast-moving beams of electrons. Wavelengths are of the order of 10^{-8} mm (0.1 Å), and, as a result, magnifications of the order of a million times are possible without loss of resolution. Focusing is effected, not by material lenses, but by the applications of either electrostatic or magnetic fields. The use of this instrument is reserved for the examination of microfine structure of materials. It is described in some detail by Hall[5] and also by Haine[6].

5.2.2. TELESCOPES

The telescope enables one to obtain magnified images of objects at some considerable distance from the eye, and it is particularly useful for providing a visual examination of a surface otherwise inaccessible. It consists essentially of two lenses, or lens systems, called the *objective* and *eyepiece*, respectively. *Figure 5.4* shows,

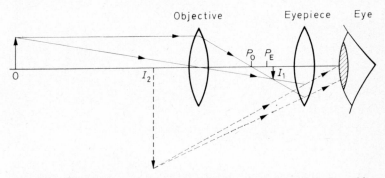

Figure 5.4. Simplified optical arrangement for telescope (not to scale). O = object under examination, I_1 = real image of O formed by objective, I_2 = virtual image of I_1 formed by eyepiece, P_0, P_E = focal points of objective and eyepiece, respectively

typically, how the image is formed. The telescope can be used in conjunction with a periscope for viewing a concealed surface but closed-circuit television is now often used for this purpose. *Figure 5.5*

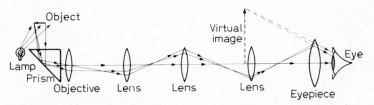

Figure 5.5. Optical arrangement for borescope (vertical scale exaggerated)

illustrates the principles of the *borescope*, a device used for inspecting the inner faces of narrow tubes and other inaccessible surfaces.

5.2.3. PROJECTORS

The reader is no doubt familiar with the projector as an instrument for viewing 'lantern slides'. It can also be used, however, to obtain a magnified image of a component on a screen, where it can be examined with comparative ease. *Figure 5.6* illustrates the basic optical principles of this device.

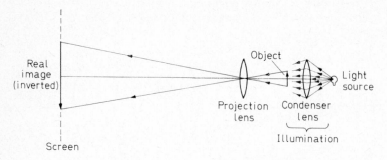

Figure 5.6. Basic optical arrangement for projector

The magnifying power of the projection lens is equal to the ratio of the image distance to the object distance. For practical reasons, it is usually undesirable for the object to be placed too close to this lens although, for any appreciable degree of magnification, the image distance must be comparatively large. *Figure 5.7* shows how a compact arrangement can be obtained. Here the beam of light is deflected several times and the final image appears on a translucent screen close to the observer. In this way, for example, continuous inspection of the profile of a screw thread can be made during the cutting process.

Where the surface of an opaque object is examined, the magnifying power of the projector is limited by the intensity of the illumination. This decreases with increase of the square of the magnifying power. Thus, for example, a magnification of 25 times corresponds to the intensity of the image being reduced to a fraction of 1/625 of that of the object, and a highly powerful source of illumination would be required.

Further details of the design of projectors are given in McMaster[2] (Vol. I).

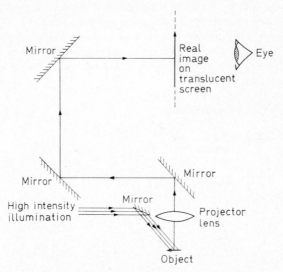

Figure 5.7. Typical layout for projector used for examining opaque objects, showing deviations of path of light rays for compactness

5.3. APPLICATIONS OF PHYSICAL OPTICS TO VISUAL METHODS OF TESTING

5.3.1. GENERAL CONSIDERATIONS

Before the fundamental principles of testing with the aid of physical optics can be understood, it is desirable to have some knowledge of the wave nature of light. A detailed account of the subject can be obtained from any standard college textbook on Physical Optics (e.g. Longhurst[7]). However, the text of the following subsection should suffice for most purposes.

5.3.2. THE WAVE NATURE OF LIGHT

Light is a form of electromagnetic radiation to which the human eye is sensitive. Other forms of electromagnetic radiation include γ-rays, x-rays, ultra-violet and infra-red radiations, radio waves, and radar waves. In a vacuum, this radiation travels at a constant speed of approximately 3×10^8 m/s (about 186,000 miles/s). In any material medium which is transparent to the radiation,

139

the speed of propagation depends on the nature of the material and, to some extent, the frequency of the source.

Electromagnetic radiation is propagated in the form of waves and is thus characterized by its *frequency f* and *wavelength* λ. In many applications, radiation occurs only at a single frequency, or, more correctly, over a narrow band of frequencies and is said to be *monochromatic* and, in the case of visible light, a single colour is observed.

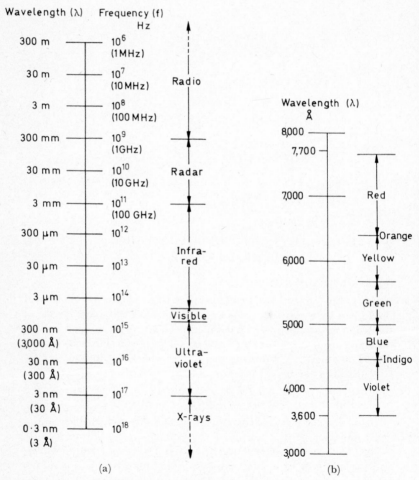

Figure 5.8. (a) Electromagnetic spectrum (logarithmic scale). (b) Visible spectrum (linear scale)

The relationship between wavelength and frequency is given by the following expression

$$c = \lambda f \qquad (5.3)$$

where c represents the speed of the waves in the material.

Figure 5.8a indicates values of λ and f for various types of electromagnetic radiation in a vacuum, for which c remains constant. This separation into the various wavelengths (and frequencies) produces what is called a *spectrum*. *Figure 5.8b* shows the visible portion of the spectrum in more detail. The wavelength varies from $7 \cdot 7 \times 10^{-4}$mm at the red end down to $3 \cdot 6 \times 10^{-4}$ mm at the violet end. It is common practice to express optical wavelengths in Angstrom units (Å), 1 Å being equal to 10^{-7} mm, and the wavelengths corresponding to the extremities of the visible part of the spectrum are 7700 Å and 3600 Å, respectively. White light, in general, produces a continuous spectrum containing radiation of all wavelengths within these limits.

Monochromatic light is generally produced by an electrical discharge tube or an arc. Most optical sources radiate monochromatic waves of a number of different wavelengths and the required wavelength is obtained by adding a suitable filter. For example, a hydrogen discharge tube radiates light of the following wavelengths: $4101 \cdot 8$ Å (violet), $4340 \cdot 4$ Å (indigo), $4861 \cdot 4$ Å (blue) and $6562 \cdot 8$ Å (red). If a blue monochromatic source of wavelength $4861 \cdot 4$ Å is required, a blue filter, opaque to all the other wavelengths listed above, must be used.

A monochromatic source of light radiates sinusoidal waves. If it were possible to measure at a given time the values of the electric field E, resulting from the radiation, at all points along a given wave at different distances d in the direction of propagation, the curve illustrated in *Figure 5.9a* would result. The waves travel at a speed c and, after a time t, the curve moves forward a distance ct to the position shown in *Figure 5.9b*. The value of E for a time t and a distance c is given by

$$E = E_0 \sin 2\pi \left(ft - d/\lambda\right) = E_0 \sin \left(\omega t - kd\right) \qquad (5.4)$$

where $k = 2\pi/\lambda$ is called the *wave number* and $\omega = 2\pi f$ the *angular frequency*. E is sometimes described as the *electric vector*. E_0 is the *amplitude* of E. The direction of E is always at right-angles to that of propagation and, for this reason, the waves are *transverse*.

141

In general, a given source of light radiates a colossal number of short trains with their electric vectors orientated at random in planes perpendicular to the direction of propagation, i.e. *unpolarized light* is radiated. However, as shown in Section 5.3.4. it is possible to produce rays in which the electric vectors are all lined up in the same direction and the light is said to be *polarized*.

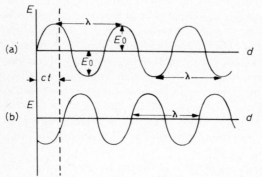

Figure 5.9. Variation of electric vector E with distance d from a given origin for part of a wave-train (a) at a given time (b) after a further time t has elapsed

It was stated earlier that when a beam of light passes through a material medium, the speed of light is reduced from the value c_0 for a vacuum to a lower value c. The ratio c_0/c is called the *refractive index n*. The magnitudes of both c and n are dependent on the nature of the medium and also on the wavelength. Thus for borosilicate crown glass, commonly used for the manufacture of optical instruments, we have that $n = 1\cdot5073$ where $\lambda = 6563$ Å and $n = 1\cdot5152$ where $\lambda = 4861$ Å.

Let a beam of monochromatic light pass from a medium 1 of refractive index n_1 into another transparent medium 2 of refractive index n_2 (see *Figure 5.10*). The relationship between the angle of incidence i and the angle of refraction r is given by

$$n_2/n_1 = c_1/c_2 = \sin i / \sin r \tag{5.6}$$

where c_1 and c_2 are the speeds of light in media 1 and 2 respectively. It should be noted that the speeds of light in air and most other gases at normal pressure are very nearly equal to the speed in a vacuum, with the result that the refractive indices of these materials are very close to unity. Equation 5.6 shows why a refracted beam of white light, which contains a continuous range of wavelengths (and thus refractive indices), is dispersed, i.e. refracted at different angles to form a spectrum. A familiar application of this phenomenon is the refracting prism used in spectroscopy (see *Figure 5.11*).

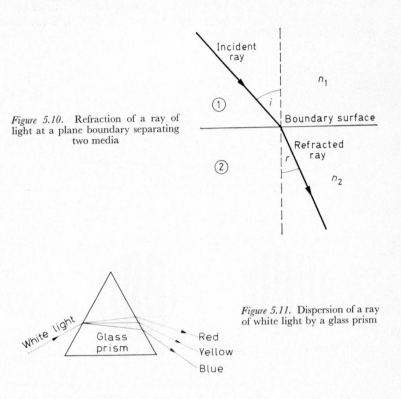

Figure 5.10. Refraction of a ray of light at a plane boundary separating two media

Figure 5.11. Dispersion of a ray of white light by a glass prism

5.3.3. INTERFEROMETERS

An important method of inspecting surfaces for flatness is the use of an optical interferometer, which can achieve an accuracy to better than one-tenth of a wavelength, typically 5×10^{-5} mm (i.e. 2×10^{-6} in). The method utilizes the interference with one another of two beams of light, a phenomenon which can take place only when the beams are coherent, i.e. the distribution of phases among the numerous wave-trains making up the beams must be the same for each. This occurs only when (*a*) the beams both originate from a common source and (*b*) the path difference does not exceed the average length of each wave-train.

Consider a beam of monochromatic light incident on a thin air wedge formed by the flat surfaces PQ and RS, inclined at a small angle to one another (see *Figure 5.12a*). Here PQ represents the lower surface of a glass plate having accurately flat and parallel faces and RS the surface to be tested for flatness. Interference of the light reflected at the surface RS with that coming directly from the source takes place at the surface PQ.

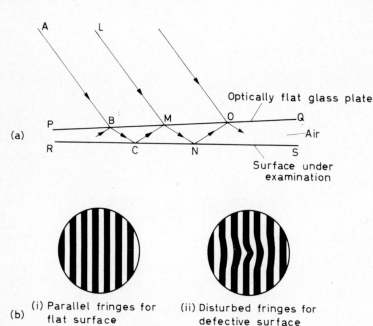

Figure 5.12. Interference at a wedge of air formed by two flat surfaces inclined at a small angle. (a) Configuration of rays of light. (b) Appearance of fringes for (i) perfectly flat lower surface and (ii) defect in flatness of lower surface

Let the rays ABCM and LM interfere in the manner shown in *Figure 5.12*. The intensity at M depends on the phase difference between the two rays. Maximum intensity is observed when the two rays are exactly in phase with one another (i.e. the phase difference is 0°, 360°, 720°, etc.) and minimum intensity when they are of opposite phase (i.e. the phase difference is 180°, 540°, 900°, etc.). The value of this phase difference depends on the path difference BCM between the two rays. If one takes into account the fact that a phase change of 180° occurs on reflection at C, it can easily be seen from equation 5.4 that the two rays are in phase when the path difference is equal to an odd number of half-wavelengths and are of opposite phase when the path difference is equal to an even number of half-wavelengths. This type of interference takes place at all points along the surface of PQ and, if a low-powered microscope is focused on to this surface, a series of bright and dark fringes parallel to the line of intersection of the two surfaces are observed, provided that the surface RS is flat (see *Figure 5.12b*). Any departure from flatness of the surface RS is indicated by the distortion of the fringes, as shown in the diagram.

The method described above is a basic one. More sophisticated techniques and their applications have been described by Tolansky[8]. Interference methods are used also in such applications as phase-contrast microscopy, photo-elasticity, and holography.

5.3.4. PHOTO-ELASTICITY

Photo-elasticity has long been applied to the investigation of stress distribution in transparent materials, such as scaled-down perspex models of engineering structures. Latterly the technique has been extended to the investigation of stresses acting in the surfaces of strained objects made of non-transparent materials which have been coated with films of transparent substances.

This technique is an application of the phenomenon of polarization, which was discussed briefly in Section 5.3.2. There are several ways of producing polarized light but the two most important are the methods of *dichroism* and *double refraction*.

Dichroism is a property exhibited by a number of substances, which can be made up into plates transparent only to light polarized in a given direction. Examples of dichroic materials are tourmaline and a specially manufactured substance known by the trade name of *Polaroid*.

Double refraction is a property possessed by anisotropic transparent materials, two of the best known being calcite and quartz. It is also a property of transparent materials, such as glass, which are normally isotropic but can be rendered anisotropic by the application of a mechanical stress. A beam of unpolarized light which passes through a doubly refracting material becomes polarized in two directions at right-angles to one another. For each of these directions of polarization there is a different speed of light and, hence, a different refractive index. Thus, for example, in *Figure 5.13* a beam of unpolarized monochromatic light is shown to be incident obliquely to a surface of a calcite prism, the optic axis of which is at right-angles to the plane of the paper. Two separate refracted rays are observed with their respective polarizations parallel and perpendicular to the optic axis, i.e. at right-angles to and in the plane of the paper. If the incident beam is of white light, the emergent beams each form separate spectra.

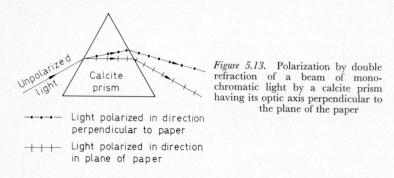

Figure 5.13. Polarization by double refraction of a beam of monochromatic light by a calcite prism having its optic axis perpendicular to the plane of the paper

—•—•— Light polarized in direction perpendicular to paper

┼┼┼┼ Light polarized in direction in plane of paper

A simple photo-elastic method of measuring surface stresses in glass by the technique of double refraction has been described by Ansevin[9]. The device used is called a *differential surface refractometer*. It consists of a rectangular glass prism (see *Figure 5.14*) placed on the surface of the test sample and illuminated obliquely by a source of monochromatic light. The material of the prism is chosen such that its refractive index n_1 is greater than the refractive index n_2 of the sample. A coupling fluid having a refractive index with a value lying between n_1 and n_2 ensures good optical contact of the prism with the sample.

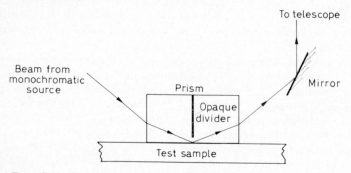

Figure 5.14. Optical arrangement for the differential surface refractometer (after Ansevin[9])

A beam of monochromatic light incident to the surface of the sample at an angle i to the normal (see *Figure 5.15*) is refracted at an angle r to the normal in accordance with equation 5.6. Because n_1

is greater than n_2, r must be greater than i and the transmitted beam is refracted away from the normal. When i is increased, r is also increased and i eventually attains a value C for which the corresponding value of r is $90°$. C is called the critical angle and, for values

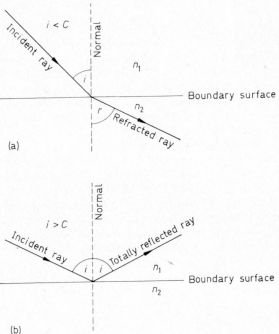

(a)

(b)

Figure 5.15. Ray diagrams illustrating total reflection

of i greater than C, no further refraction is possible and total reflection takes place at the surface. It can easily be shown that

$$\sin C = n_2/n_1 \tag{5.7}$$

With the arrangement shown in *Figure 5.14*, the field of view through the telescope is divided into two portions separated by a well-defined line. One portion is illuminated by that part of the light incident at angles greater than C and totally reflected. The other portion is dark because the remainder of the light incident, at angles less than C, is not reflected. To prevent any unwanted light reaching the telescope the prism contains an opaque divider, as shown in the diagram.

147

When the sample is stressed mechanically, it becomes doubly refracting and the light passing through it is polarized in two directions at right-angles to one another. It can be shown that the surface stress T is related to the refractive indices n_2' and n_2'' of the material, for each of the polarizations, by

$$T = k(n_2' - n_2'') \qquad (5.8)$$

Here k is a constant having a value dependent on the nature of the material. There are thus two critical angles C' and C'' corresponding to n_2' and n_2'', respectively, and two distinct lines of demarcation are observed through the telescope. If a suitably calibrated graticule is placed in the focal plane of the eyepiece of the telescope, the value of T can be determined directly from the distance of separation of the two lines.

In general, the techniques used in photo-elasticity are far more complicated than the method just described and the observation of interference fringes is usually involved. Details of some of these are to be found in volume II of McMaster[2]. In one of the more advanced techniques the surface under examination is coated with a thin film of transparent plastic material and the stress distribution in this film corresponds closely to that in the surface of the loaded sample. In this way, non-transparent objects can be tested. The thickness of this film can vary between 0·005 and 0·050 in (0·13 and 1·3 mm), depending on the degree of sensitivity required for the measurements.

5.3.5. HOLOGRAPHY

Holography is the name given to a method of obtaining an accurate three-dimensional image of a given object. The process is carried out in two stages. First a permanent record in the form of a two-dimensional interference pattern is obtained on a photographic plate by means of a laser beam. The image is then obtained from the record, again using a laser.

The method of recording is illustrated in the ray diagram shown in *Figure 5.16*. A monochromatic beam of light from the laser is incident partly to a mirror and partly to the object under test. Interference takes place, at the photographic plate, of light reflected from the mirror with light scattered by the object. The plate is developed and fixed and a complicated system of fringes, called a *hologram*, is thus recorded on its surface. In view of the long path difference between the reflected and scattered beams, one cannot use an ordinary monochromatic source. This is because of the lack of coherence as a

148

result of the path difference being very much greater than the average length of the wave-trains in the beams. The laser, however, radiates coherent wave-trains of more than a metre in length.

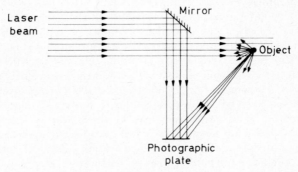

Figure 5.16. Ray diagram illustrating the production of a hologram

Figure 5.17 illustrates how the image can be obtained from the hologram, which diffracts the laser beam. Two three-dimensional images are observed. One of these, which is virtual, can be viewed with the unaided eye or some optical instrument without any risk

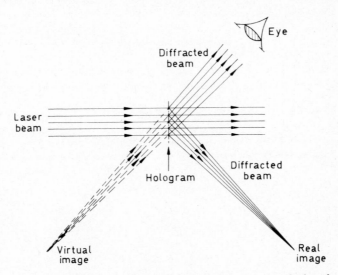

Figure 5.17. Diffraction of a laser beam by a hologram to form both real and virtual images

149

of damage to the eye because the observer does not look directly at the laser source. The other image, which is real, can be recorded on a photographic plate. In this way a picture having a high definition and free from aberrations can be obtained without the use of a camera.

The technique of holography is still in its early stages of development but already it has a number of important applications. One can envisage its usefulness in the non-destructive testing of the surfaces of highly complicated precision components without the disadvantages of having to use a high-power microscope. In a single operation, a hologram can provide a record of the image of an entire surface which can be readily compared with that of a standard defect-free surface. The comparison may be made with a high-speed automatic scanning device which can be adjusted to detect flaws having a magnitude greater than a strictly specified tolerance. Further details of the technique of holography are provided by Stroke[10].

REFERENCES

1. HEAVENS, O. S., *Progress in Non-Destructive Testing* (Eds. Stanford, E. G. and Fearon, J. H.) Heywood, London, 1961, Vol. 3, p. 3.

2. McMASTER, R. C. *Nondestructive Testing Handbook* (Vols. I and II), Ronald, New York, 1959.

3. MARTIN, L. C. *Technical Optics* (Vol. II), Pitman, London, 1950.

4. HABELL, K. J. and COX, A. *Engineering Optics*, Pitman, London, 1948.

5. HALL, C. E. *An Introduction to Electron Microscopy* (2nd Edn.), McGraw-Hill, New York, 1966.

6. HAINE, M. E. *The Electron Microscope*, Spon, London, 1965.

7. LONGHURST, R. S. *Geometrical and Physical Optics*, Longmans, London, 1957.

8. TOLANSKY, S. *Multiple Beam Interferometry of Surfaces and Films*, Oxford University Press, Oxford, 1948.

9. ANSEVIN, R. W. *Mater. Eval.* 1967, **25**, 58.

10. STROKE, G. W. *An Introduction to Coherent Optics and Holography*, Academic Press, New York, 1966.

6

PENETRANT METHODS

6.1. GENERAL CONSIDERATIONS

THE penetrant method of non-destructive testing is employed in the inspection of non-porous materials for detecting those discontinuities which are difficult to observe by normal visual examination. It utilizes a penetrating liquid which is applied to the surfaces to be tested and absorbed by any discontinuities present. The excess penetrant is removed after a suitable period of time and a developing agent applied to the surfaces under test. The developer draws the penetrant from the discontinuity thereby revealing its location.

The penetrant method in its earliest form was in existence prior to World War II. It was commonly known as the *oil and chalk process*, and consisted of a heavy lubricating oil diluted with paraffin or a light mineral oil giving a penetrant of medium viscosity. The part to be examined for surface discontinuities was immersed in this liquid for a reasonable time, after which it was removed and the excess penetrant swabbed off with dry rags or cotton waste moistened with paraffin. When the surfaces were clean and dry, a mixture of chalk and alcohol was applied with a brush. This coating dried rapidly giving a white surface. Any discontinuities would have been revealed by brown stains on the white background, due to seepage.

The method did not usually reveal small discontinuities and was largely superseded by the magnetic particle method (see Chapters 3 and 4), where applicable. Variations of the above system, where the oil was heated to a temperature of 80°C, were used with limited success.

One of the first penetrant processes to utilize a fluorescent material and ultra-violet radiation was used in Great Britain during World War II and to a large extent replaced the conventional oil and chalk process. The parts to be tested were immersed for 30 minutes in trichlorethylene to which had been added a fluorescent compound anthracene.

In order to remove the excess fluorescent material and provide a relatively clean background, the parts were washed in carbon tetrachloride after removal from the penetrant and allowed to dry prior to viewing with ultra-violet light. This process was used mainly

151

for the inspection of light alloy components, and was eventually replaced by a system using a coloured dye in a powdered form mixed with the developer powder.

The components were immersed in white spirit for 30 minutes, wiped dry and sprayed with the red dye/chalk mixture. Discontinuities were shown as vivid red indications whilst the background was a pale pink. This system was reasonably effective and, if applied with care, fairly small imperfections could be revealed.

With the introduction of the American 'Zyglo' fluorescent penetrant into Britain at the end of World War II, a complete revolution in penetrant testing techniques took place. This particular process had been in use in America since 1942 and was the result of intensive work carried out by R. C. Switzer and the Magnaflux Corporation, who conducted development work to ensure a practical system for industrial use.

At about the same time work was being carried out to develop dye penetrants, and these generally came into industrial use in 1945. Further development work by the Magnaflux Corporation, using fluorescent penetrants, produced the *post-emulsification method* which was patented in 1957. Improvements to all the types of penetrant methods in use have been made by companies in America and Europe since the process became an industrial success.

In 1950 Brent Chemicals Ltd. (now Ardrox Ltd.) introduced a penetrant dye process designated 'Ardrox 996'. In 1956 this company obtained a licence agreement from the De Havilland Aircraft Company to manufacture and market a water-washable fluorescent penetrant designated 'Ardrox 970'. Following this, Brent Chemicals Ltd. introduced post-emulsifiable fluorescent penetrants and obtained the patent rights of the 'cascade' phenomenon[1]. 'Cascading' works on the principle that when a particular fluorescent dye does not emit light of the desired colour or intensity at the black light (ultra-violet) peak wavelength of 3650 Ångstroms, another dye may be receptive to the emitted light of the first dye and, in consequence, re-emit light of the desirable wavelength and colour. A combination of two or more such dyes produces a penetrant of greater brilliance than a single dye.

Recent developments in the penetrant field have been an improved type of 'Ardrox 996' a red dye penetrant which becomes fluorescent when viewed under black light. In collaboration with Rolls-Royce, Aero Division, they have also developed a penetrant removal method, utilizing a foam bath consisting of an aqueous emulsifier. This has resulted in the detection of extremely fine discontinuities, as seen in certain gas-turbine components.

Another major British contributor to penetrant testing is the Manchester Oil Refinery Ltd., now Castrol Industrial, who have developed and marketed a full range of penetrant materials for a variety of applications.

Since the introduction of penetrant testing, considerable developments have been carried out by various companies into the equipment used for the process, which ranges from hand operated units to fully automatic systems.

6.2. METHODS OF TESTING

To obtain optimum results from penetrant testing, the following sequence of operation must be carried out.

(1) *Pre-cleaning*

The parts to be examined must be thoroughly cleaned and, in particular, the discontinuities must be free from contaminants such as water, oil, etc. One of the most effective cleaning methods is the use of trichlorethylene vapour. Another is by immersion or swabbing in a suitable solvent. In general, abrasive methods should not be used unless followed by a light acid etch as these tend to peen over the surface of the discontinuity, thereby stopping or reducing the entry of the penetrants.

(2) *Application of the penetrant*

This may be achieved by dipping the component in a bath of penetrant liquid or by spraying or brushing. A minimum penetrant contact time must be allowed after application. This is usually 20 to 30 minutes, but it may be extended in certain circumstances.

(3) *Removal of excess penetrant*

If a water washable penetrant has been used, it is sufficient to wash the part with water. If a post-emulsifier penetrant has been applied, washing takes place after application of the emulsifier. As its name suggests, a solvent-clean penetrant is removed by using a solvent. The cleaning operation is required to remove the excess penetrant from the part in order to give a maximum defect indication of the defect against a clean background.

(4) *Drying*

To obtain well defined indications, the component should be dried with a cold or warm air blast at low pressure prior to the development stage, unless the application of an aqueous wet developer is required, in which case the drying stage can be omitted.

(5) *Developing*

A coating of developer is applied to the part under examination by one of the following methods. It can be sprayed on in powder form, placed in a powder 'storm' cabinet after the surfaces have been dried, or dipped into an aqueous solution. The developer, which is usually a white powder, acts as a blotting agent to enhance the tendency of the penetrant to exude from the discontinuity. For optimum results, a thin even coating should form as a background to the penetrant indications. It is usual to allow approximately 20 minutes for completion of development prior to viewing the component.

(6) *Inspection*

After the appropriate developing time has elapsed, inspection can be started. Viewing should take place in an area supplied with good intensity white light, if a dye penetrant has been used, and in a darkened area provided with an ultra-violet light of a specified wavelength if a fluorescent penetrant has been used. The action of the ultra-violet light is to cause the penetrant to fluoresce, thereby indicating discontinuities readily.

6.3. APPLICATIONS AND LIMITATIONS OF THE PENETRANT METHOD

The penetrant method, although having a wide application, should always be regarded as an aid to inspection and not always the final answer to what type of discontinuity is present or indeed that the indication is deleterious. In many cases the indication should be marked for a later visual assessment. In order to obtain the optimum results from penetrant testing, a full understanding of the capabilities and limitations of the method should be appreciated.

The process is capable of detecting discontinuities penetrating to the surface of the material under test. In metals, these are usually cracks, laps, seams, porosity, etc. Penetrants have also been widely used in testing vessels and pipes, etc. for leaks. It is essential that the defects are free from dirt and grease, otherwise the penetrant will not enter the opening. A crack which is shallow and wide is more difficult to reveal because it is relatively easy to wash the penetrant out. Since the introduction of post-emulsifier methods this limitation has been greatly reduced. Probably the largest use of penetrant testing is for light alloy material examination. The penetrant method is equally applicable to magnetic materials, but generally the magnetic particle test is preferred because it is well established and it will detect subcutaneous discontinuities, defects filled with oxide or carbon and defects covered by paint films. It is generally cheaper and also less cleaning is required than for the penetrant method. One advantage the penetrant method has over the magnetic particle method is, that when large numbers of small complex parts are required to be inspected, it is usually quicker because handling becomes less of a problem.

Penetrant testing is complementary to the other methods of non-destructive testing. For instance, although radiography will locate internal discontinuities, it is not reliable in locating cracks, unlike the penetrant technique. Penetrant testing has a significant advantage over other non-destructive testing methods, with the possible exception of magnetic particle inspection, in that a part can be tested over its complete surface in a relatively short time, irrespective of shape and size and defect orientation, thereby reducing costs. Difficulties can often arise in this respect with the use of ultrasonic and eddy current methods of testing. Penetrant testing is very reliable in the detection of fatigue cracks which occur during the service life of materials.

The penetrant method is often considered simple by the uninitiated and adequate training is usually lacking. Given reasonable equipment and a full understanding of the methods, capabilities and limitations by both the operator and supervision, the process is extremely effective and profitable. Equipment can range from simple hand operated units, obtained at a relatively low cost, to the much more expensive fully automatic units. To decide on the purchase of equipment, account must be taken of the numbers and shapes of parts to be tested, as well as the type of penetrant to be used. Staff should have adequate training for the particular process under their control.

Compromise must be exercised in the choice of a penetrant with regard to its nature and properties and, in this respect, a vast amount of experience has been accumulated over the years by the manufacturers of penetrant materials.

6.4. THE PROPERTIES OF PENETRANTS

The desirable features of an ideal penetrant are that it must be

(a) able to enter very fine discontinuities,

(b) not easily removable from wide discontinuities until required,

(c) not easily removable from shallow discontinuities until required, (*d*) easily removable from the surface of the component,

(*e*) capable of being easily drawn from discontinuities by the developer,

(*f*) able to spread as a very fine film,

(*g*) in possession of intense colour or fluorescence,

(*h*) able to retain its colour or fluorescence when exposed to heat, white light or ultra-violet light of the correct wavelength,

(*i*) inert to the materials including containers with which it comes into contact,

(*j*) odourless,

(*k*) non-flammable,

(*l*) stable when stored and in use,

(*m*) non-toxic, and

(*n*) cheap.

It is fairly obvious that the physical properties possessed by a given material are most unlikely to provide all the characteristics mentioned above for an ideal penetrant. No one physical property more than another determines which material or combination of materials makes a suitable penetrant. The properties, apart from availability and cost, which must be considered for the choice of a penetrant, include viscosity, surface tension, wetting ability, density/specific gravity, volatility, flash point, chemical inertness, solubility, solvent ability, emulsifiability, ability to spread, tolerance for contaminants, toxicity, odour and the danger of skin irritation.

(1) *Viscosity*

This controls the rate at which a penetrant will enter a discontinuity. If a penetrant is viscous, it offers more resistance to entering an opening and, to reduce penetrant losses to a minimum due to 'drag out', the draining period has to be increased, thus increasing the time of the process. This feature is particularly serious when using the post-emulsification process, as extensive contamination of the emulsifier will occur thus resulting in increased costs. If the viscosity of the penetrant is too low, it will drain rapidly from the part, leaving insufficient penetrant available to enter the discontinuity, and it will also be susceptible to easy removal by the washing process.

(2) *Surface tension*

Generally speaking a penetrant should have a high surface tension for a high degree of penetration.

(3) *Wetting ability*

Good wetting ability is essential for high penetrating characteristics. Wetting ability is determined by the contact angle between the penetrant and the surface under test, at the point of contact, the smaller the contact angle, the better the wetting ability. Penetration ability is determined by a high value of $T \cos \alpha$, where T is the surface tension and α the contact angle for a particular penetrant and a given surface. The value of α is dependent on the surface roughness and cleanliness of the material. A high degree of wetting is essential for good penetrating ability but it can only be specified for the particular surface involved, i.e. a clean smooth surface of one material might be satisfactory for a given penetrant but unsatisfactory where that surface is in a rough dirty condition, for the same penetrant.

(4) *Density/Specific gravity*

This property has no direct influence on the penetrating ability of the penetrant but, if the specific gravity is less than unity, water contamination will be less of a hazard because the water will settle at the bottom of the tank.

(5) *Volatility*

Generally speaking, a penetrant should be non-volatile because, otherwise, loss takes place by evaporation and, more important, it tends to dry too rapidly on the surface or in the discontinuity, thus providing difficulties in removal or in fault finding.

(6) *Flash point*

A satisfactory penetrant must have a high flash point, to lessen the risk of an explosive mixture forming with the air above the liquid surface. The usual minimum flash point for modern penetrants is 130°F (55°C).

(7) *Chemical inertness*

Penetrants should be as inert and non-corrosive as possible with regard to their containers and to the materials under test. Oil base penetrants usually meet these requirements but special formulations may be needed for testing nickel alloys, rubber or plastic parts. Similarly a special penetrant may be required when testing parts which will eventually come in contact with liquid oxygen.

(8) *Solubility*

A penetrant must have good solubility in order to be removed in a reasonable manner and time from the surface under test.

(9) *Solvent ability*

This must be extremely high because penetrants contain fluorescent or coloured dyes, the latter usually at a high concentration. The penetrant must be capable of dissolving the dyes at ambient temperature and retaining them in solution at possible low temperatures during storage or transportation.

(10) *Emulsifiability*

It is very important that this feature of a penetrant is an optimum, otherwise extreme difficulty will be encountered in its successful removal. This applies both to water washable penetrants, in which the emulsifier is incorporated, and to post-emulsifier penetrants for which there are separate emulsifiers.

(11) *Ability to spread*

This property is required in order that small amounts of penetrants will be able to emerge from a discontinuity to form a 'seeable' indication with the aid of a developer.

(12) *Tolerance for contamination*

Although normal precautions should be taken to avoid it, contamination of a penetrant with water, grease, oil etc., does take place due to inadequate degreasing of the parts and carelessness on the part of the operator. Furthermore, care must be taken not to site the penetrant plant adjacent to processes liable to produce contamination. Although manufacturers of penetrant materials provide a formula which is capable of tolerating a degree of contamination, this should never be used as an excuse for allowing contamination to take place.

(13) *Toxicity, odour and skin irritation*

Generally these hazards are not present in modern penetrants to such a degree that the health of the operator is threatened.

It can be seen, from the above-listed requirements and the foregoing remarks regarding the physical properties, that the manufacture of a suitable penetrant is a difficult task. Since penetrant testing was first introduced, both manufacturers and users have sought to find ways of measuring their properties and characteristics such as penetrability, sensitivity, contrast and 'seeability'.

It is generally considered that capillary action is the basic phenomenon responsible for the penetration of a liquid into a discontinuity. Various attempts have been made to measure the 'fineness' of discontinuity with regard to the possibility of penetration. One of these, reported in 1951, was a glass plate experiment[2] in which two glass plates 10 in (250 mm) long, $1\frac{1}{2}$ in (35 mm) wide and $\frac{3}{8}$ in (9·5 mm) thick were clamped together. When sufficiently tight, it was possible to observe interference fringes extending outwards from the clamped areas when they were viewed under sodium light, wavelength 5896 Å. The distance between any two consecutive fringes represents a change of separation of one half of the wavelength of the light used to produce the fringes. It was found, when using this method, that the penetrant liquid would enter a separation in the order of 5 micro-inches (0·1 μm).

Other methods used in an attempt to determine the penetrating ability of liquids are the testing of fine cracks produced in hard nickel plating on steel strip and the testing of cracks in a fatigued metal specimen. These attempts do enable penetrants to be tested in a reasonable manner, but obviously more information will be available when it is possible to find a method to reproduce natural cracks. Unfortunately, this has not been achieved to date.

6.5. SENSITIVITY

Since the introduction of penetrant examination, various experiments have been conducted and ideas thus formulated to measure sensitivity. Sensitivity can be defined as the ability of any one penetrant to reveal a particular type of discontinuity in a given material. This is related to fine or wide discontinuities which contain depth or which are shallow in nature. Factors affecting sensitivity are the ability of the penetrant to enter the discontinuity and removal of the penetrant from the surface of the component without any significant removal from the defect. In addition, the penetrant must have the ability to exude from the discontinuity, with the aid of a developer, and to form an indication which is readily visible.

Various attempts at manufacturing devices or test blocks in order to obtain a measure of sensitivity and to compare penetrants have been tried over the years. One of these devices designed in 1957 by the A.I.D. Laboratories, Harefield, was a demountable block[3]. The construction is illustrated in *Figure 6.1*. Areas identified A, B, C and D on one barrel section are relieved to the extent of 0·0001 in (2·5 μm), 0·0002 in (5 μm), 0·0003 in (7·5 μm), and 0·0004 in (10 μm) respectively and are at intervals of 90° to each other around the periphery. The mating faces of both barrel sections were previously lapped.

When in use, the lapped faces are tightened together by applying a torque loading of 200 lb ft (15 kg m) to the nut. To achieve a discontinuity and to prevent damage to the lapped faces, a lead/tin/antimony foil 0·0007 in (17·5 μm) in thickness is inserted, prior to

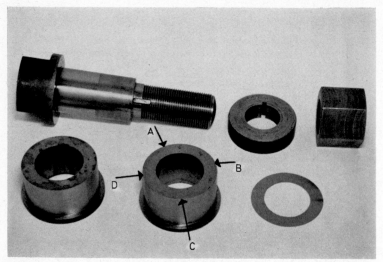

Figure 6.1. Demountable block, disassembled (*By courtesy of A.I.D., Harefield*)

applying the torque loading. The device is then subjected to penetrant examination in the normal manner and inspected for defect indication at the four points around the periphery (see *Figure 6.2*). The test piece suffers the following disadvantages:

(*a*) The discontinuity is not a natural one, particularly as the lapped faces bear little or no resemblance to the texture of a natural defect.

(*b*) The machined surface is favourable to the use of solvent developers because dry developers do not work as well on a smooth surface. It follows therefore, that the effect of developers on the sensitivity of flaw detection is not controlled.

(*c*) The finest discontinuity obtainable is of the order of 100 micro-inches (2·5 μm) whereas it is known from other experiments that widths of 10 micro-inches (0·25 μm) or less are required to differentiate penetrating abilities of reliable penetrants.

The advantages of using the test piece for showing artificial defects are that, depending on the torque loading, a varying width of defect can be produced and the defects can be thoroughly cleaned between tests.

Another useful way of obtaining samples for penetrant comparisons is to prepare fatigue specimens. For medium and large cracks a standard Wohler specimen can be used but, for best results, where extremely fine cracks are required, a cantilever specimen excited in its fundamental mode of vibration is used. The cracks can be detected during a fatigue test by a change in the frequency of this mode. This method has the advantage of producing natural cracks but, once again, it is impossible to obtain reproducibility.

Figure 6.2. Demountable block, assembled (*By courtesy of A.I.D., Harefield*)

An extremely useful test-piece for the evaluation of penetrants is the special aluminium test block which is recommended by the Air Registration Board[4]. The block is machined from 24.ST. aluminium to the following dimensions, $4\frac{1}{2}$ in (112·5 mm) long, 2 in (50 mm) wide and $\frac{3}{8}$ in (9 mm) thick. A slot is machined down the centre of the face in order to provide separate test surfaces for the comparison of different penetrants.

Cracks are produced as an overall pattern in both faces of the block by heating to 525°C in not less than 4 minutes and quenching in cold water. The cracks produced cover a wide range in width and depth and can therefore be used for comparison of penetrants.

After use the blocks can be cleaned by vapour degreasing and heating to 125°C. If they are not required immediately after cleaning, the blocks can be stored in acetone or toluene.

The upper test block illustrated in *Figure 6.3* shows a direct comparison between a water washable penetrant on the left-hand side and a post-emulsifiable penetrant on the right-hand side of a slot

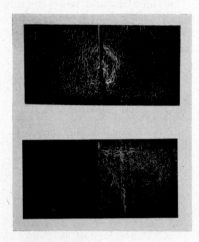

Figure 6.3. Comparison between water washable and post-emulsifiable penetrant (upper specimen). Comparison between dye and dye fluorescent penetrant (lower specimen) (*By courtesy of Messrs. Rolls-Royce Ltd., Small Engine Division*)

cut in the surface. The post-emulsifier penetrant is seen to reveal finer defects. The lower test block shows the advantage of the use of a dye penetrant which also fluoresces under black light (right-hand side). The left-hand side of the block was processed using a water washable dye penetrant. This revealed approximately the

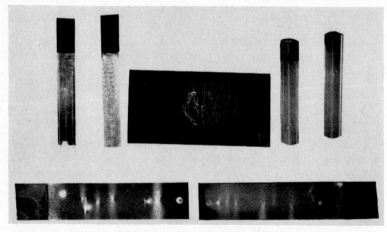

Figure 6.4. Comparison between post emulsifiable (specimens on left) and water washable penetrant (*By courtesy of Messrs. Rolls-Royce Ltd., Small Engine Division*)

same number of cracks as the dye-fluorescent type of penetrant but the cracks were only visible in white light.

Another type of test-piece which is widely used is a steel bar or a strip of material which has been chromium plated to a thickness of approx. 0·005 in (125 μm). The round specimens are ground to produce cracks, while the strips are indented with Brinell hardness equipment to produce discontinuities. The major advantage with this type of specimen is that the defects can be easily cleaned by vapour degreasing.

The specimens shown on the left-hand side of *Figure 6.4* (including the Air Registration Board block) were processed using a post-emulsifiable penetrant. The remainder were processed using a water washable penetrant. When considering the use of dye or fluorescent penetrants, some consideration should be given to their relative contrast and 'seeability' factors. Contrast is defined as the relative amount of reflected and/or emitted light between an indication and its background. 'Seeability' is defined in terms of the amount of light reflected or emitted by an indication. This feature involves colour, ambient light level and contrast. It is an established fact that dye penetrant will give a maximum ratio of 9:1, while fluorescent penetrants will give a light to darkness ratio of approximately 300:1. As can be seen from the chart of human eye perception (*Figure 6.5*), dye penetrants must be viewed in good

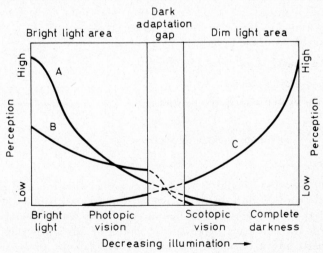

Figure 6.5. Chart of perception by the human eye. A. Perception of contrast. B. Perception of colour. C. Perception of small light sources
(*By courtesy of Magnaflux Corporation*)

white light so that the ability of the eye to distinguish contrast and colour is at its maximum. When viewing fluorescent penetrant indications, the eye must be allowed to adapt itself for at least 10 minutes to the darkened condition prior to viewing. Providing this is carried out properly, the 'seeability' of minute fluorescent light spots is very high because the eye is immediately drawn to a source of light in a dark background. Furthermore, halation occurs, making the size of the light source appear larger. Additionally, the maximum colour response of the eye is in the yellow/green region, which is the colour emitted by modern penetrants. The 'seeability' of fluorescent indications is therefore higher than dye penetrants.

6.6. DEVELOPERS

A developer is a coating applied to the test surface to reveal the penetrant indication more easily. A good developer has a blotting action which tends to draw more penetrant liquid from a discontinuity and to provide a background which makes the indication readily visible with good contrast. The overall sensitivity of the method is thus increased and the time taken for the indications to become visible is diminished. Developers must be (*a*) absorptive to obtain maximum blotting action, (*b*) fine grained to give sharply defined indications, (*c*) able to mask the background of the part and improve contrast, (*d*) applicable in thin even uniform coatings, (*e*) easily wetted by the penetrant, (*f*) not itself fluorescent and (*g*) easily removable from the surface.

The four types of developers are

    (1) dry powders,

    (2) water suspensible powders,

    (3) water-soluble developer, and

    (4) solvent suspensible developers.

(1) *Dry powders*

Dry powders should be light and fluffy with the ability to cling to surfaces thus giving an overall fine film without excessive build-up, which might mask fine defects. They should neither be hygroscopic nor toxic. The powder should be applied by dipping the parts into a container, by spraying with a gun, or by using a dust storm cabinet in which the powder is agitated by compressed air.

(2) *Water suspensible powders*

This type of developer is widely used because, when large numbers of small parts are required to be processed, they can be placed into baskets and immediately after the washing sequence they are dipped into the wet developer which gives an even coating to all the parts.

No drying stage is required after removal of the penetrant, but drying is applied after the developer stage. It is advantageous to heat the developer tank which, of course, accelerates drying.

(3) *Water soluble developers*

Difficulties have always been experienced in formulating suitable materials which are soluble in water. Thus the water-soluble developer has never really become a commercial proposition.

(4) *Solvent suspensible developers*

This type of developer is applied by pressurized containers or spray guns and has the advantage of giving a thin even coating which dries rapidly. Developer powder should always be removed after final inspection of the part. This can easily be carried out by washing with warm water.

It is essential to use the correct type of developer for the given penetrant process and type of work surface under test. As a general guide, wet developers should be used on very smooth surfaces and dry developers on rougher surfaces. Other considerations on the choice of developer are that wet developers are useful for large quantities of small parts, but these can give excessive build-up where there are severe changes in section. Solvent developers are more satisfactory on tight defects than on wide shallow defects due to the solvent action of the penetrant.

6.7. PENETRANT PROCESSES

6.7.1. FLUORESCENT PENETRANTS

6.7.1.1. *Water-washable Process*

Before any penetrant examination is carried out, the parts must be thoroughly cleaned. Failure to carry out this operation properly will result in discontinuities not being revealed, false indications, and the production of a dirty background, resulting in low overall contrast. All loose dirt, rust, scale, etc. can be removed by wire brushing. It is not generally recommended to clean mechanically using a blasting process such as vapour or grit blasting as it has been proved quite conclusively that a peening effect which closes openings of any discontinuities present takes place[5]. If for any reason a type of blasting has to be used, it should be followed by a light acid etch, followed by washing with warm water and vapour degreasing. Experiments have shown that this procedure is normally satisfactory.

Figure 6.6 shows fatigue cracks in three alloys, heat resisting nickel base, 12 per cent chromium steel and case hardening steel. The defects were revealed by a water washable fluorescent penetrant.

Figure 6.6. Fatigue specimens showing cracks (*By courtesy of Messrs Rolls-Royce Ltd., Small Engine Division*)

Figure 6.7 shows the same specimens after barrelling, the dark areas being those of corrosion. It can be seen that the surface treatment has reduced the ability of the penetrant to reveal the defects. Removal of oil, grease etc. is most satisfactorily achieved by vapour degreasing which in addition, ensures freedom from water.

Where there are deposits of carbon on the surface due to service environment, special industrial cleaners are used to faciliate removal. The parts are then washed in water and degreased by vapour. If it is required to remove paint, the parts should again be washed thoroughly in water after removal and the process is followed by vapour degreasing. It is generally considered that if a normal type of paint has been used, it will crack in sympathy with the parent material during service, and it is therefore unnecessary to remove it prior to penetrant examination.

Ultrasonic cleaning of components prior to penetrant examination is also useful, because it tends to break up and loosen oxide, carbon

166

etc. and thus may reveal part, if not all, of the discontinuity. It is *most* important to ensure the absence of chromates, chromic acid, etc. from any part to be tested, as these tend to kill the fluorescence of the penetrant.

Figure 6.7. Fatigue specimens as in *Figure 6.6.* after barrelling (*By courtesy of Messrs. Rolls-Royce Ltd., Small Engine Division*)

When the cleaning process has been satisfactorily carried out, the penetrant can be applied by various means such as dipping, brushing and spraying. The particular method employed is usually decided as a matter of convenience. Large numbers of very small components can be placed in a wire basket and immersed in a tank containing the penetrant. Medium sized parts can be placed in a tank individually whilst, for larger components, the penetrant can be applied by brushing and spraying. When the dipping procedure is used, care must be taken to ensure that no air pockets are present in recesses and blind holes, and that the volume of penetrant is sufficient to cover the work. Whichever method is used to apply the penetrant, the critical feature is the penetrant contact time. If this is too short, insufficient time will be available for the penetrant to enter the discontinuity and, if too long, there will be a tendency for the penetrant to dry on the surface and become difficult to remove by washing, thereby giving an unsatisfactory background. From these considerations it is normal to stipulate a penetrant contact time of 20 to 30 minutes, which of course includes immersion and draining times, if applicable. *Figure 6.8* shows a typical area for penetrant application, water wash, drying and developing.

Various procedures often suggested to aid the penetration of the penetrant include to date, heating of the penetrant fluid, heating of the parts prior to immersion in the penetrant, the use of pressure and/or vacuum, and vibration. Unfortunately none of the methods has been completely effective.

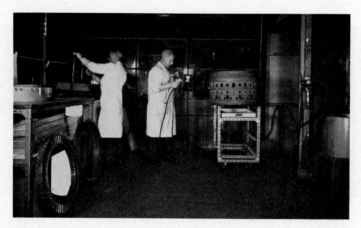

Figure 6.8. Typical penetrant testing area (*By courtesy of Messrs. Rolls-Royce Ltd., Small Engine Division*)

The next stage of the process is removal of the penetrant by water washing. It is essential to employ the correct washing technique, otherwise the efficiency of flaw detection is greatly reduced. The most common method used is the hosepipe with a nozzle fixed to its end, so that the jet of water, composed of coarse droplets, impinges on the surface to be washed. The pressure of water should be about 30–40 lb in^{-2} (2–3 $\times$ 10^{-3} kg mm^{-2}) and the temperature in the range of 15 to 40°C. If higher pressures are used there is a danger of removing the penetrant from the defect. It is essential that the penetrant is washed away quickly and efficiently otherwise, once emulsification has taken place, some dye will adhere to the surface of the part and it will be extremely difficult to remove, causing a poor background.

It is preferable to carry out the washing operation under ultra-violet light, so that the washing can be terminated as soon as the part is clean, thus avoiding unnecessary removal of penetrant from defects. Washing should always be carried out by starting at the bottom of the component, to avoid water running down, and by separating oil from emulsifier before the spray reaches the area concerned. Care should also be exercised to ensure cleanliness of baskets, if used, and also the surrounding area of the wash tank, otherwise false indications will result from fluorescent 'pick up'. If, for any reason a spray wash cannot be carried out, local areas can be washed by swabbing. When the washing process is complete, the components are dried prior to developing, unless a wet developer is to be used.

Drying can be successfully accomplished by using compressed air, but re-circulating hot air dryers operating in the temperature range 65 to 85°C are generally considered more efficient because, when the parts are heated, the penetrant starts to exude from the discontinuity as a result of the heat absorbed by the surface of the part. In addition, the definition of the indication is greatly improved due to the volatilization of some penetrant, which in turn causes the higher viscosity constituents to contain the indication.

Either dry powder or a wet developer (water suspensible) may be used for the water washable process. If the latter is to be used, the drying stage can be omitted and the parts are immersed in the developer, followed by draining and drying as described above. When applying dry powder it is essential for the parts to be thoroughly dry, otherwise excessive patches of powder will appear. The time allowed for development should be approximately 20 minutes.

After development the components are ready for inspection. The inspection area should be darkened because fluorescent indications emit visible light. A completely darkened area is not essential; in fact it is preferable to have a low intensity red or amber light for background illumination. Adequate ventilation should also be supplied.

Where ultra-violet illumination is used, it should consist of a tubular 'flood light', placed down low to avoid the operator's eyes, together with a high intensity 'spot lamp' for local examination of inaccessible areas. In order to cut out background ultra-violet light which can be reflected from the component, it is advisable for the operator to wear goggles with a yellow filter, this increases the viewing contrast and avoids fluorescing of the eyeballs. *Figure 6.9* shows a typical penetrant viewing area.

It is extremely important for cleanliness to be observed in the viewing booth, otherwise isolated spots of fluorescent penetrant will adhere to benches, curtains, lamps and the operator's hands. These in turn may well be transmitted to the part being examined causing false indications.

The final stage after viewing of the parts and assessment of indications should be to remove the developer powder.

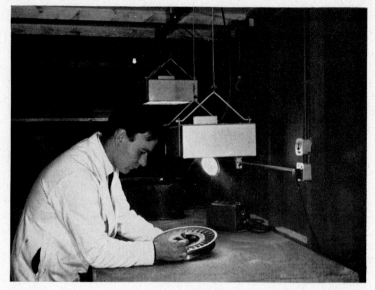

Figure 6.9. Penetrant viewing booth (*By courtesy of Messrs. Rolls-Royce Ltd., Small Engine Division*)

6.7.1.2. *The Post-Emulsifier Process*

The difference between the post-emulsifier process and the water washable process is that the application of the emulsifier is a separate stage. This enables the degree of washability to be controlled, depending upon the time of contact of the emulsifier with the penetrant. In *Figure 6.10* the upper test block shows the effect of over-emulsification on the left-hand side, compared with normal emulsification on the right-hand side. The lower test block compares under-emulsification on the left-hand side with normal emulsification on the right-hand side. The intensely fluorescent areas are finger marks caused by careless handling.

Other than this, the process stages are the same as for water washable penetrants, i.e. pre-cleaning, application of penetrant, removal of excess penetrant by water washing, drying, application of developer and inspection for discontinuities using ultra-violet light.

The advantages of the post-emulsification method are as follows:

(1) Because the degree of washability can be controlled, open shallow defects can be located.

170

(2) Maximum penetrability is obtained due to the absence of emulsifier in the penetrant which, in turn, achieves higher sensitivity for fine defects.

(3) Owing to the absence of the emulsifier in the penetrant liquid, the concentration of fluorescent dyes can be increased, thus resulting in high brilliance.

(4) The absence of emulsifier allows the penetrant liquid to enter discontinuities faster, thus reducing the penetrant contact time.

(5) Acids and chromates only react with fluorescent dyes in the presence of water. The penetrant process does not contain or tolerate water and this hazard is therefore considerably reduced.

(6) Parts can be reprocessed several times, with consistently good results, because there is no emulsifier present to contaminate the discontinuity when vapour degreasing is carried out as in the water washable process.

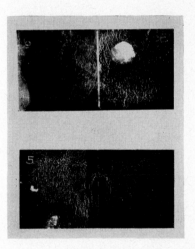

Figure 6.10. Post emulsification process.

Top left, over emulsified
Top right, normal emulsification
Bottom left, under emulsification
Bottom right, normal emulsification

The disadvantages of the method are:

(1) It is a two step process prior to washing.

(2) Careful timing of the emulsification stage is required to obtain optimum sensitivity.

(3) When processing complex parts, care must be exercised to ensure emulsification of the complete surface area, otherwise

difficulties in the removal of the penetrant at the washing stage will result.

(4) The cost is higher than that for the water washable process.

6.7.2. DYE PENETRANTS

Dye penetrants were introduced shortly after washable fluorescent penetrants in an attempt to improve sensitivity. As they contained no emulsifiers they were to some extent able to detect finer discontinuities.

The advantages of dye penetrants are:

(1) They are simple to use.

(2) They are extremely portable, requiring no ultra-violet light for inspection.

(3) They can be supplied in kits, utilizing pressurized containers, at relatively low cost.

(4) They are particularly useful for field work where mains services may not be available.

The disadvantages are:

(1) The maximum contrast ratio possible is 9:1 (fluorescent penetrants achieve 300:1).

(2) A thicker developer layer is required to obtain maximum contrast.

(3) They are generally more costly.

(4) The problem of staining clothes, equipment etc. must be considered.

It is an established fact that modern post-emulsifiable penetrants are more sensitive than dye penetrants because the fluorescent system is superior in 'seeability'. However, dye penetrants still occupy a useful place in the field of testing, particularly when lower sensitivity of flaw detection is adequate.

Dye penetrants are now available in three versions, namely:

(1) Solvent-clean type.

(2) Water wash type.

(3) Post-emulsifiable type.

The processing sequence for the solvent-clean type of penetrant is as follows:

(a) Pre-cleaning of the surface; as described for fluorescent penetrants.

(b) Application of dye penetrant; as described for fluorescent penetrants.

(c) Removal of excess penetrant, by wiping the surface of the part with a clean dry cloth or absorbent paper damped with solvent remover.

(d) If (c) is not sufficient, the process is repeated until a clean background is obtained. Care should be taken to avoid excessive use of solvent remover, otherwise penetrant liquid may be removed from any discontinuities present. The surface of the part is finally wiped dry with a clean cloth or absorbent paper.

(e) Application of developer; the developer is applied by spraying and is of the type which dries rapidly thus making it easy to detect if any parts have not been covered.

(f) Inspection for flaws, which should be carried out in daylight or good artificial illumination approximately 5 minutes after application of the developer.

The processing sequence for the water wash type of penetrant is the same as for the solvent clean type except for the penetrant removal stage and the drying stage. The penetrant is removed by spray rinsing with water (as for fluorescent penetrants) and the component is dried by using clean compressed air or by placing it in an air circulating oven.

The procedure for processing the post-emulsifiable penetrant is again the same except for the application of a detergent type penetrant remover followed by water wash and drying.

6.8. REMOVAL OF PENETRANTS

Having completed the process to the emulsification stage a penetrant remover is then used. The two main types of remover are designated *hydrophilic* and *lipophilic*. To appreciate the differences in characteristics between these, it is necessary to understand the basic theory of penetrant removers. In the post-emulsifiable method, as its name implies, the penetrant is neither emulsifiable nor directly water washable until it comes into contact with a penetrant remover. The contact time between the penetrant and remover will control the sensitivity of the process, which is critical for maximum sensitivity. If the surface film of penetrant can be kept uniformly even and thin, only brief contact times are necessary, which improves the sensitivity of the process.

173

A lipophilic penetrant remover is one which depends on mutual solubility between it and the penetrant and therefore contact times must be kept to a minimum to avoid dilution of the penetrant in discontinuities, with subsequent reduction in flaw detection sensitivity. This type of remover is sensitive to water contamination.

The advantage of a hydrophilic remover is its insolubility in the penetrant. This, together with high water tolerance, enables excess penetrant to be removed by water washing prior to application of the remover. Thus a thin uniform film of penetrant can be obtained, which reduces the contact time. It has been shown that higher levels of sensitivity can be achieved when using an aqueous penetrant remover instead of a penetrant compatible remover[6].

Penetrant removers must be formulated to obtain the correct hydrophilic/lipophilic (H/L) balance. For penetrant compatible removers, by increasing the H/L balance, more efficient removal of excess penetrant can be achieved but if the balance is excessively low, a heavy background fluorescence will remain. Hydrophilic removers are normally prepared as concentrates and can be diluted by up to 20 times their volume by water, thus making their use very economical and allowing the removal of excess penetrant by water washing prior to emulsification.

It has also been shown that extremely fine fatigue cracks can be located by reducing background fluorescence. To achieve this, a gentle stream of air is passed through a diluted solution of a hydrophilic remover, which produces a continuously moving column of foam[6].

The sequence of operations for this process is as follows:

The sequence:
(1) Pre-clean and degrease.
(2) Apply penetrant (contact time 20 minutes).
(3) Air/water wash.
(4) Apply foam.
(5) Air/water wash to remove foam.
(6) Dry.
(7) Develop.
(8) Inspect.

6.9. AUTOMATED PENETRANT PLANT

With the introduction of post-emulsifiable penetrants, the process can be fully automated. The advantages to be gained over hand processing are as follows:

6.9. AUTOMATED PENETRANT PLANT

(1) Standardization of the processing cycles, i.e. penetrant contact time, emulsification time, etc.

(2) A greater throughput of work per unit time.

(3) Saving in total processing costs.

(4) Optimum sensitivity for the particular penetrant process being used.

Both dye and fluorescent post-emulsifiable systems are in use with automatic equipment but the latter, due to its increased sensitivity, is normally preferred. *Figure 6.11.* shows a front elevation of a typical automatic penetrant unit which is in use for the examination of gas turbine aero-engine blades. The components are loaded by hand into wire baskets which are then attached to the conveyer rail of the unit. The baskets are indexed by a series of steel pawls attached to a bar supported by rollers and connected to an air cylinder, the operation of which results in all the work loads being transferred to the next stage.

Vertical movement of the baskets at the dipping stages is controlled by a vertical air cylinder. The pawls, which are used to transfer the baskets from stage to stage, return to their original position ready for the next indexing operation. In the particular unit being described, the work loads are indexed along at 5 minute intervals. The time of immersion at each dipping station is also preset.

After vapour degreasing has been carried out, the basket containing the components is attached to the rail (stage 1). The basket is then indexed to stage 2 at which it is immersed in the penetrant liquid for 1 minute, followed by 4 minutes drain and transfer to stage 3 at which there is a 5 minutes drain. This is followed by transfer to stage 4 for a further 5 minutes drain, and to stage 5 for a final drain of 5 minutes. The total penetrant contact time is thus 20 minutes. The contact time can be varied according to the number of stages built into the equipment. The baskets then move on to stage 6 at which there is a cold water spray rinse consisting of 3 minutes in the rinse tank, with vertical agitation at 10 second intervals, followed by 2 minutes drain and transfer to stage 7, the penetrant removal process. The basket waits for a 3 minute period over the tank followed by immersion for 1 minute with vertical agitation at 10 second intervals, followed by 1 minute drain and transfer to stage 8, the second cold water spray rinse, which is identical to stage 6. After rinsing, the basket arrives at stage 9 at which there is immersion for 10 seconds in an aqueous developer, followed by draining and warm air drying for the balance of 5 minutes. The basket is then transferred to a further drying stage

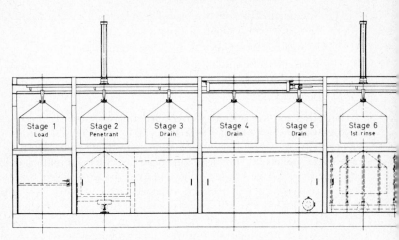

Figure 6.11. Automatic processing unit, front elevation (*By courtesy of Ardrox Ltd.*)

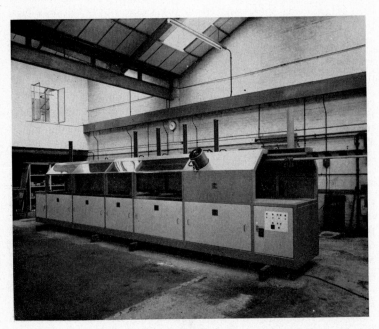

Figure 6.12. Automatic processing unit, general view (*By courtesy of Ardrox Ltd.*)

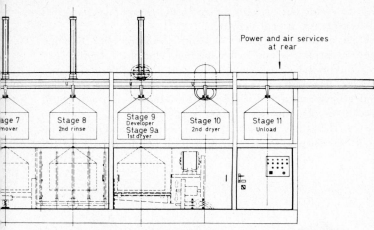

Power and air services
at rear

| age 7 mover | Stage 8 2nd rinse | Stage 9 Developer Stage 9a 1st dryer | Stage 10 2nd dryer | Stage 11 Unload |

Figure 6.11 (contd.)

10 for a further 5 minutes of warm air drying and then to stage 11 for cooling and unloading. The components are then ready for final inspection and unloading. When a dry developer is used, the components enter the drying section after stage 8 and are then placed for 2 minutes under low-pressure air jets to remove excess water, followed by 10 minutes of warm air drying and finally, 5 minutes in a dry powder dust storm cabinet.

Figure 6.12 shows a general view of the equipment used for this process. It is 30 ft 10 in (9·40 m) long, 5 ft 3 in (1·58 m) wide (allowing for all services) and 7 ft 10 in (2·39 m) overall height. The electrical control panel is recessed into the equipment at one end and the compartment can be locked once the settings are made.

Automatic penetrant units can be designed to suit a variety of types of work and have already proved their value beyond all possible doubt in the aero-engine industry.

6.10. THE USE OF ULTRA-VIOLET LIGHT

Ultra-violet light is that part of the electro-magnetic spectrum in the wavelength range approximately 2000 to 4000 Ångstroms. For fluorescent penetrant inspection, the wavelength used is in the range 3200 to 4000 Ångstroms with a peak wavelength of 3650 Ångstroms. Radiation of these wavelengths, near the visible end of the spectrum, is commonly termed 'black light'. The dyes best suited for the formulation of fluorescent penetrants react most strongly when energized with black light of 3650 Ångstroms wavelength. They

177

usually emit visible light within the green to yellow range, which is most favourable for optical detection.

There are a number of types of lamps available for producing black light satisfactorily and by far the most efficient is the mercury vapour arc type, which is available both as a portable hand lamp or a tubular 'flood lamp'. These lamps are fitted with glass filters which effectively remove all visible light produced and the harmful radiation of wavelengths below 3000 Ångstroms.

Figure 6.13 illustrates a typical 100 watt portable hand lamp. The black light intensity is liable to deterioration due to various factors

Figure 6.13. Ultra-violet portable handlamp (*By courtesy of Magnaflux Corporation*)

and, therefore, regular measurements should be made to maintain optimum output. Any reduction or variation in intensity of lamps would be due to the following causes:

(*a*) Variation in bulbs during manufacture.

(*b*) Intensity variation with applied voltage.

(*c*) Fall off in intensity with increasing age of the bulb.

(*d*) Accumulation of dirt on the bulb, filter and reflector.

To some extent the variations can be reduced, e.g. by fitting voltage stabilization units, by avoiding starting the lamp more than necessary in a working day, and by cleaning lamps regularly.

It is still desirable, however, to measure the intensity of a lamp, and the following procedure has been proved satisfactory for use on the shop floor, necessitating the minimum of apparatus and skilled personnel. The equipment consists of a viewing hood (see *Figure 6.14*) constructed of metal with the internal surfaces blackened to stop unwanted reflections. The fluorescent screen consists of zinc cadmium sulphide phosphors which fluoresce when irradiated by

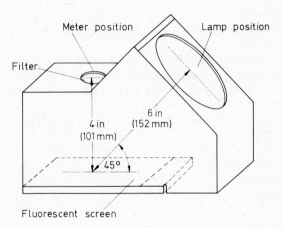

Figure 6.14. Ultra-violet viewing hood (*By courtesy of Messrs. Rolls-Royce Ltd., Small Engine Division*)

black light from the lamp which is placed at 45° to the screen and 6 in (15 cm) distant from it. The visible light emitted from the screen is measured on a light meter or photometer. A minus blue filter such as an Ilford micro 108 yellow filter is placed between the light meter or photometer to remove any reflected black light. When not in use the fluorescent screen is kept in a light proof casette to avoid deterioration. With proper care, the screen will maintain its efficiency for about 10 years.

To measure loss of fluorescent from the screen, a small section can be covered with adhesive tape and subsequently used as a standard. The measured intensity using the above method should not be allowed to fall below 10 ft candles.

REFERENCES

1. BRITISH PATENT 844540.

2. BETZ, C. E. *Principles of Penetrants*, Magnaflux Corporation, Illinois, 1963.

3. ROBERTS, R. W. and DUDBRIDGE, A. D. *A.I.D. Technical Note, N.D.T. 938*, August 1957.
4. AIR REGISTRATION BOARD CIRCULAR BL/10-9. *Performance Testing of Penetrant Testing Materials*, April 1965.
5. MCFAUL, H. J. *Mater. Eval.* 1965, **23** (12), 577.
6. BIRLEY, R. E., HYAM, N. H., JACOBSEN, K. M. and TEBBENHAM, J. *The Brit. J. Non-Destruct. Test.* 1967, **9**, 90.

For Further Reading

Non-Destructive Testing Handbook. McMaster, R. C. (Volume 2), Ronald, New York, 1959.

Air Registration Board Circular BL/8-7. *Fluorescent Penetrant Process.*

Air Registration Board Circular BL/8-2. *Dye Penetrant Process.*

British Standard 3683: Part 4: 1963. *Glossary of terms used in Non-Destructive Testing, Penetrant Flaw Detection.*

British Standard 3889: Part 3A: 1965. *Penetrant Testing of Ferrous Pipes and Tubes.*

American Military Specifications
    MIL-1-251356.
    MIL-1-19684.
    MIL-1-9445.A.
    MIL-1-25105.
    MIL-1-25106.
    MIL-1-6866.B.

APPENDIX

QUALITY CONTROL OF SMALL COMPONENTS USING FÖRSTER'S MAGNATEST Q EQUIPMENT

A.1. SCOPE

FÖRSTER'S Magnatest Q equipment (see Section 2.4.8) is normally employed for the checking of steel components which are reasonably large and heavy, thus giving a strong signal which is not affected by positional or other errors. It is, however, feasible to use it to measure quite small differences of permeability for components down to one gramme in weight, provided the equipment is suitabley jigged and adjusted to minimize the effects of the different operating variations which can always occur. For large components, however, they are swamped by the application of a strong signal. These variations, unless controlled, can render impossible the sorting of small parts, for which the signals are weak.

A.2. ANTICIPATED SENSITIVITY

The following tests can be made by careful attention to detail:

(*a*) Sorting 12 per cent chromium steel jet engine turbine blades for conformity to a required hardened and tempered hardness value, with a maximum error equivalent to 3 HV on components down to 1 gramme in weight.

(*b*) Sorting bolts for hardness as small as 4 BA with a maximum error equivalent to less than one ton tensile strength.

(*c*) Measuring decarburization down to levels of 0·001 in (0·025 mm) deep on bolts down to 1/4 in (6 mm) diameter.

(*d*) Checking core condition on fully heat treated and nitrided parts on items as small as 5 grammes in weight.

(*e*) Checking case depth and/or case hardness on surface hardened parts down to 1/2 (12 mm) diameter.

Problems such as these may not come within the reader's province, but a knowledge of the factors to be controlled can often enable the borderline application to be handled easily and can equally be used to improve results when testing larger objects.

A.3. FACTORS AFFECTING SENSITIVITY

Machine Factors (See Section A.4)

1. Basic sensitivity.
2. Coil size.
3. Power Input.
4. Sine curve function selected ($\int E$, E or D indications)
5. Portion of sine curve selected.
6. Sensitivity switch position.
7. Variation of voltage into machine.

Feed Factors (*see Section* A.5)

1. Speed of belt (not applicable to hand feeding).
2. Position in coil.
3. Orientation in coil.

Component Factors (see Section A.6)

1. Material.
2. Total weight.
3. Shape, i.e. diameter to length ratio in particular.
4. Uniformity.

A.4. CONTROLLING MACHINE SENSITIVITY FACTORS

A.4.1. Basic Sensitivity

There are a number of small circuit adjustments which can be made by the manufacturers (Forster Instruments) and these should be incorporated if the machine is not of the latest type. Standard machines operate at 50 Hz mains frequency and prove adequate for the testing of small components, as already stated. For equipment specifically required to check skin effects, there is an advantage in using frequencies of 100 Hz or higher. Frequencies of 25 Hz are less suitable.

A.4.2. Coil Size

As a rough guide, halving the coil size doubles the sensitivity obtainable. Consequently the coil size should be as small as possible. A device to improve the effectiveness of a given coil is discussed later (see Section A.7.5).

A.4.3. POWER INPUT

Increasing the power input to the primary windings of the testing coil will increase the sensitivity in proportion. However, there are two disadvantages in using more power. The first is that a point can be reached where the component is held stationary in a coil by the flux-field and then slips on the conveyer. With hand feeding, which is not ideal for precision work, this snag does not occur but a point can be reached, using slightly more power, where the component is over-saturated and a corresponding increase in sensitivity does not occur. At this point there is a reduction of sorting ability.

The second disadvantage is of major importance when checking skin effects. With low power inputs, the effect of the skin layer is the major factor in the response but with high power input, the degree of penetration into the work is increased. Thus the lowest feasible power should be used for skin testing and the highest for checking core or overall states, so as to swamp any difference in the skin layer. The sensitivity setting should be adjusted for compensation.

A.4.4. SINE FUNCTION SELECTED

Unless the component has a very simple form, e.g. a bar of regular cross-section, the use of the $\int E$ function, which gives a smooth sine curve is necessary for small parts.

A.4.5. PORTION OF SINE CURVE USED

All Magnatest Q instruments have a 360° switch which enables the operator to centre any part of the curve. In any sorting problem, components of the similar composition, although of different hardness, will give their maximum peak at the same point of the sine curve. This point should be selected for hardness sorting. However components of varying composition, even of the same hardness, will show maximum response at different points on the sine curve. For analysis, a portion of the curve should be selected to give maximum differential between the different compositions (i.e. analysis groups).

It is obvious, therefore, that batches of work containing different analyses, as well as varying hardnesses, must be sorted twice at different switch position settings. It is important to note that a change in the power input will shift the response and require re-setting.

A.4.6. SENSITIVITY SWITCH POSITION

The sensitivity switches (coarse and fine) will give varying sensitivities in stages from 0·1 to 10·0. Because the total sensitivity is the product of the power input and the sensitivity setting, it is important to set the desired power level first to obtain skin effects or deep seated responses, as desired, and then to increase the sensitivity enough to give a satisfactory differentiation between the 'acceptable' and the 'reject'. With visual readings, this can be as low as 1 mm but, with chute gates fitted, the response must be generally of the order of 4 mm difference to give satisfactory sorting. Over-sensitivity should be avoided because it can provide too great a response and adversely affect automatic sorting.

A.4.7. VOLTAGE INPUT

In a busy works, the varying load through a working day can seriously alter machine settings by causing voltage drops up to 10 per cent. It is essential for critical work to fit a voltage stabilizer, obtainable from the manufacturers, to the machine.

A.5. CONTROLLING FEED FACTORS

Hand feeding is not ideal for precision sorting because the very accurate positioning necessary for consistency can only be obtained with precision jigs. A given component, which is either offset by a fraction of an inch or out of line by a few degrees, will provide a different signal and the error can swamp the small difference which is to be detected. By using a conveyer, together with a lining up device, it is always possible to position components precisely (see Section A.7.3). Also, since the component passes right through the coil, the optimum response position along the coil axis is always reached.

The belt speed has a bearing on the automatic sorting gate controls. A higher speed can cause a borderline soft component to fail to trigger the gate. Minor adjustments of the belt speed can be used for fine limit setting.

The position of the component in the coil is *very important*. Each successive component must ideally be placed in the centre of the coil axis and on insertion the response must reach its maximum midway through the bore. With hand feeding, this can be detected visually from the height of the sine curve but belt feeding is much to be preferred because the position of the component can be made precise.

The orientation of a component in the coil is also vital. The component, wherever possible, should be placed lengthwise in the coils since a large length/diameter ratio of the part will improve the response. An error of 5 degrees in alignment will have a noticeable effect on the result. Thus some sort of jigging becomes essential (see Section A.7).

A.6. COMPONENT FACTORS AND THEIR EFFECT

A.6.1. MATERIAL

Constructional steels vary in their basic magnetic permeabilities. A high permeability such as that possessed by soft iron laminations will give a bigger response than a mild steel component. Hardened and quenched components will give a lower response than unhardened components, even if the Vickers hardness due to cold work is the same. Martensitic stainless steels provide even smaller responses. This factor can limit the minimum size of the components to be tested although, in practice, the response is sufficient unless components weighing less than 1 gramme are to be sorted.

A.6.2. WEIGHT

Heavier components will give a bigger response than lighter components and thus require lower power and sensitivity settings. In practice, however, it has been found that weight variations not greater than 5 per cent, within a batch, have a less marked effect on the response than they suggest and small differences in analysis type or hardness will still be detectable.

A.6.3. SHAPE

The main variation in components, which affects the response, is the length/diameter ratio. Very short components have a self-demagnetizing effect which counteracts the response. It is sometimes possible to control this by testing them at an early machining stage, when the length is greater, or by deliberately planning the machining stages so as to manufacture the components in joined pairs. For components of complex form, irregularities in the contour of the sine curve occur but, generally, sorting is not hindered, provided that a suitable portion of the sine curve is centered with the 360° switch.

A.6.4. UNIFORMITY

Components with minor cracks, soft or hard spots, or small irregularities have surprisingly little effect on the response. However,

185

it is obviously unwise to attempt to sort components with major form variations because these variations can mask the desired sorting difference.

A.7. MACHINE MODIFICATIONS TO MINIMIZE UNWANTED VARIABLES

A.7.1.

Having selected the smallest possible coil, the most important requirement is to ensure that each successive component is positioned precisely in the same position. With hand feeding, this often necessitates the use of simple jigs which fit into the test coil. The jigs can be made of wood or plastic but not magnetic steel. In any application, it is ideal to provide stops on the jigs to locate the part centrally on the mid-axis.

Figure A.1. Wooden centring device for balancing coil (*By courtesy of Messrs. Rolls-Royce Ltd., Small Engine Division*)

A.7.2.

There is no need to position the balancing component in the comparison coil and, in fact, it is not essential even to use a similar component, provided that the magnetic effect of the balancing component is approximately the same, because adjustments to the instrument can ensure true balance. It is, however, advantageous to place the component centrally within the coil with the possible help of simple wooden or plastic sleeves of varying thickness (see *Figure A.1*).

A.7.3.

With conveyer feeding, a very worthwhile addition is an adjustable alignment device such as shown in *Figure A.2*. This is made of perspex and austenitic steel, fixed with brass studs. The gap can be

Figure A.2. Alignment device for conveyer coil (*By courtesy of Messrs. Rolls-Royce Ltd., Small Engine Division*)

set precisely to suit any object and the wings extending into the coil ensures precise positioning during the test (see *Figure A.7*). The components can then be placed on the conveyer approximately in the correct position and be aligned and centred automatically.

187

A.7.4.

When a conveyer chute is fitted, fragile parts may be damaged by dropping on the metal faced conveyer or by rocketing off the shute ends. *Figure A.3* shows (*a*) a rubber sheet at the chute throat, (*b*) a rubber guard at the chute base, (*c*) a convenient stand for reception of the sorted components.

Figure A.3. Chute fitting to minimize damage (*By courtesy of Messrs. Rolls-Royce Ltd., Small Engine Division*)

A.7.5.

To improve the responses of coils, outer mild steel jackets are placed over each coil to provide flux paths through the air and thus to complete the flux flow through the components under test. The jacket must be tailored to fit a coil as shown in *Figure A.4*. A jacketed coil can give 50 per cent greater height of response on the screen than an unjacketed coil for the same setting.

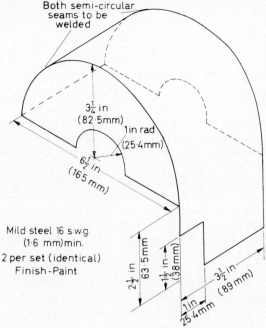

Both semi-circular
seams to be
welded

$3\frac{1}{4}$ in
(82·5mm)

1in rad
(25·4mm)

$6\frac{1}{2}$ in
(165 mm)

Mild steel 16 s.w.g.
(1·6 mm)min.
2 per set (identical)
Finish-Paint

$2\frac{1}{2}$ in
63·5mm

$1\frac{1}{2}$ in
(38 mm)

$3\frac{1}{2}$ in
(89 mm)

1in
25·4mm

Figure A.4. Jacket to improve coil response. This size of
concentrator is a snug fit for a 35 mm coil. The dimensions should be adjusted for coils of other sizes (*By courtesy
of Messrs. Rolls-Royce Ltd., Small Engine Division*)

A.8. SETTING-UP PROCEDURES

A.8.1. GENERAL

It is essential in all sorting procedures to have reliable standards.
When a new sorting application is initiated, these standards are
normally not available. However, a preliminary sorting through say
24 specimens in a batch will show differences in the sine curve
responses and the highest and lowest can be picked. Checks for
Vickers hardnesses, analyses, etc., can be made to establish two
known samples.

For batches of work which possess more than one variable, the
establishment of suitable known samples can be a complex procedure.
The variables in a batch of uniform size parts may consist of one or
more of the following factors:

(*a*) General hardness.

(*b*) Analysis (i.e. chemical composition).

189

(c) Heat treatment state, e.g. 'as forged' or 'heat treated'.

(d) Grain size.

(e) Case-hardness on case-hardened parts.

(f) Core-hardness on case-hardened parts.

(g) Case-depth on case-hardened parts.

It is unlikely that more than two of these variables occur in a batch for sorting but this possibility must be borne in mind.

In the common case where the parts in the batch are of different hardness and of mixed analysis, one would make a rough sorting to separate different analysis types and then re-sort each analysis type into separate part-batches. The same form of procedure must apply where any two or more variables occur. For this reason, it is worth while to ensure, as far as possible, that successive batches of test components are not mixed prior to Magnatest-Q testing. The following setting-up procedures each deal with checking a single variable and several initial sortings may have to be made to reduce each part-batch to a single variable.

A.8.2. Checking Components for Overall Hardness

As an example, the components are assumed to be $\frac{1}{2}$ in (12 mm) dia. bolts 3 in (75 mm) long. The following procedure must be carried out.

(a) Switch on the machine and allow it to warm up for at least two minutes prior to testing.

(b) Select the smallest pair of coils which will accommodate the bolts.

(c) Select one coil to be used as the comparison coil and mark it 'C'.

(d) Provisionally set the dials as follows:
Current setting (11) 0·6
Current adjuster (14) to read 50 amperes.
Sensitivity coarse (12) 6
Sensitivity fine 5
Function (16) ∫E
360° Switch 180°

(e) With empty coils adjust the signal to a straight line.

(f) Increase the coarse sensitivity control progressively to 10 and re-adjust the signal to obtain a straight line.

(g) Place a bolt in each coil. The balance coil component should be placed as near as possible to the axis by using a wooden sleeve or other means. The component in the test coil should be placed in a suitable centring jig for hand feeding. With conveyer systems, centring will be automatic, but the conveyer should remain stationary.

(h) Adjust each component separately to give maximum response. This is evidenced by the trace movement rising to a peak and then falling.

Note 1. The 360° switch must be adjusted at the same time so that the maximum peak is coincident with the centre line of the screen.

Note 2. The balance coil response should first appear from the top of the screen and the test coil response should appear from the bottom. If the converse is observed, the 360° switch should be adjusted by a full 180°.

Note 3. If the signal movement is too strong or too weak, the sensitivity coarse control should be altered to compensate for this.

(i) Re-adjust the trace to a straight line. If necessary the sensitivity may be temporarily reduced. *From now on, do not disturb the load in the comparison coil.*

(j) Replace the bolt from the test coil with a third bolt and note signal divergence.

(k) Repeat (j) with at least 20 components and select the two with the greatest signal divergence.

Note 1. Soft components will give a higher reading than harder components.

If the signal response is too large or too small, adjust the coarse and fine sensitivity to suit.

Note 2. At this stage, do not alter the power.

(l) Hardness-test the two bolts with the greatest signal divergence to establish a temporary sorting band.

(m) Re-test the two selected bolts to give a satisfactory screen height difference.

It is preferable to increase the power input to the fringe of saturation so that the lowest feasible sensitivity can be used, thus reducing the effect of any skin variation.

191

Note. It is very important to re-set the degree switch to re-centre the trace should the power be altered.

(*n*) Finally adjust the fine sensitivity so that the mm scale on the screen bears a relation to the hardness difference for the two samples, e.g. Set a 280 HV component to zero and a 300 HV to $+2$ mm. Then a 290 HV component will automatically read $+1$ mm and a 270 HV component -1 mm.

(*o*) With hand sorting, sort the rest of the batch, establishing go and no-go limits with bolts of borderline hardness, as found.

(*p*) With conveyer sorting, run the belt at a suitable speed and note the modified response of the two standards. The limit gate settings and, if necessary, the sensitivity can then be altered. For minor adjustments to the sorting limit, the belt speed can be adjusted. Faster speeds will accept a borderline component; slower speeds will reject a borderline component (see *Table A.1*).

Table A.1. Magnatest 'Q' Conveyer Advance (*By courtesy of Messrs. Rolls-Royce Ltd., Small Engine Division*).

The conveyer when running shows a *delay* in response according to its speed.

The quoted advance on the stationary settings should be made anti-clockwise on each threshold control to allow for this delay.

Sensitivity used	Conveyer Speed			
	10	15	20	25
11 (max.)	Nil	0·6 (3)	1·6 (9)	2·2 (12)
10·5	Nil	0·6 (3)	1·5 (8)	2·1 (11)
10	Nil	0·5 (3)	1·3 (7)	2·0 (11)
9·5	Nil	0·5 (3)	1·1 (5)	1·6 (9)
9	Nil	0·4 (2)	0·8 (4)	1·2 (6)
8	Nil	0·4 (2)	0·8 (4)	1·1 (5)
7 & below	(Not determined)			

(Bracketed figures indicate the approximate response equivalent in hardness HV for a specific component weighing 5 g).

A.8.3. CHECKING COMPONENTS (E.G. DECARBURIZED BOLTS) FOR SKIN HARDNESS

The procedure as in Section 8.2 should be followed with the following differences:

(*a*) Select two bolts showing equal high readings and two showing equal low readings. The necessary micro-examination to establish depth of decarburization can then be made on one of each pair.

(b) Use *Maximum Sensitivity* and the lowest power consistent with adequate sorting to confine the testing to the outer skin of each component.

(c) Because there is a certain amount of offset of the trace with varying decarburization levels, carefully adjust the 360° switch.

A.8.4. CHECKING COMPONENTS FOR ANALYSIS OR GRAIN SIZE DIFFERENCE

In this procedure the height of the response is not a sorting parameter. Two different analyses or grain size will give their maximum response at offset positions on the screen. For hand sorting, it is usually best to centre the trace, using the 360° switch, with one known analysis type and to sort into one or more separate batches any components offset from this position. It is quite possible to sort simultaneously as many as seven different types, provided that the general shape of the curve is recorded (see Section A.9 and *Figure A.6*).

With conveyer sorting, it is only possible to sort into three batches. The position of the trace is adjusted with the 360° switch until, at the centre position, the maximum difference between the traces occurs.

A.8.5. CHECKING FOR CASE DEPTH

The procedure here depends on using *Sensitivity* below the maximum (say 7 or 8 on *Coarse*) and a suitable power to obtain a penetration roughly matching the usual case depth. Both trace height variation and amount of trace offset will vary, and the visual sorting method using recorded shapes of the screen traces is probably the most suitable.

A.9. RECORDING MACHINE SETTING AND STANDARD

To simplify a repeat test, it is essential to record the machine settings and helpful if a collection of known standard specimens can be accumulated. A form found to be useful is based on one supplied by the manufacturers, (see *Figure A.5*). This can be prepared, if needed, on translucent paper which can be clipped on to the oscilloscope screen for the sine curve shapes to be recorded with a grease pencil trace. The figure shows a simple record for top and bottom

QUALITY CONTROL OF SMALL COMPONENTS

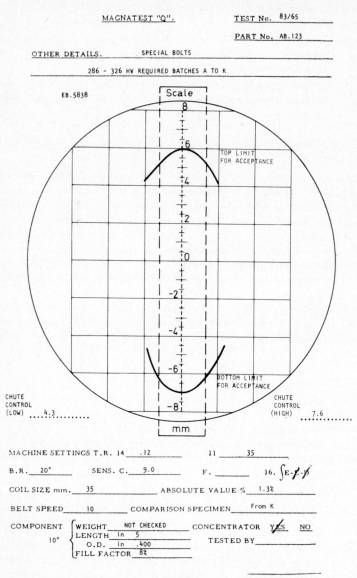

MAGNATEST "Q".   TEST No. 83/65

PART No. AB.123

OTHER DETAILS.   SPECIAL BOLTS

286 - 326 HV REQUIRED BATCHES A TO K

EB.5838

Scale

8
6   TOP LIMIT
    FOR ACCEPTANCE
4
2
0
-2
-4
-6
    BOTTOM LIMIT
    FOR ACCEPTANCE
CHUTE   CHUTE
CONTROL   CONTROL
(LOW) ....4.3.........   (HIGH)   7.6
-8   ....

mm

MACHINE SETTINGS T.R. 14 ___.12___   11 _____35_____

B.R. __20°___   SENS. C. __9.0___   F. _____   16. $\int E\text{-}\mathcal{I}\text{-}\mathcal{D}$

COIL SIZE mm. ___35___   ABSOLUTE VALUE % __1.3%__

BELT SPEED _____10_____   COMPARISON SPECIMEN __From K__

COMPONENT ⎰WEIGHT___NOT CHECKED___   CONCENTRATOR YES   NO
   10°   ⎱LENGTH __in 5___
        O.D. __in .400___   TESTED BY_____
        FILL FACTOR __8%___

NOTE: RE-ADJUST CHUTE SETTINGS IF NEEDED FOR FURTHER BATCHES

Figure A.5.   Screen record showing high/low limits for hardness sorting (*By courtesy of Messrs. Rolls-Royce Ltd., Small Engine Division*)

194

A.9. RECORDING MACHINE SETTING AND STANDARD

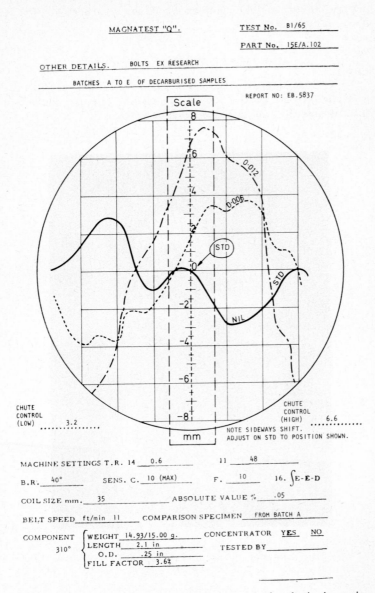

MAGNATEST "Q".

OTHER DETAILS.   BOLTS EX RESEARCH

BATCHES A TO E OF DECARBURISED SAMPLES

TEST No. B1/65

PART No. 15E/A.102

REPORT NO: EB.5837

Scale

mm

CHUTE CONTROL (LOW) .....3.2.....

CHUTE CONTROL (HIGH) .....6.6.....

NOTE SIDEWAYS SHIFT.
ADJUST ON STD TO POSITION SHOWN.

MACHINE SETTINGS T.R. 14 ___0.6___   11 ___48___

B.R. __40°__   SENS. C. _10 (MAX)_   F. __10__   16. $\int$E-E-D

COIL SIZE mm. ___35___   ABSOLUTE VALUE % ___.05___

BELT SPEED __ft/min 11__   COMPARISON SPECIMEN __FROM BATCH A__

COMPONENT   { WEIGHT __14.93/15.00 g.__   CONCENTRATOR YES NO
310°   { LENGTH ___2.1 in___   TESTED BY_____
               { O.D. ___.25 in___
               { FILL FACTOR __3.6%__

Figure A.6. Screen record showing varying sine curves for decarburization sorting
(*By courtesy of Messrs. Rolls-Royce Ltd., Small Engine Division*)

195

hardness limits; a more complex record for decarburized stock is shown in *Figure A.6.*

A.10. GENERAL PRECAUTIONS

(1) Use the Magnatest Q on a wooden bench and take care to avoid the presence of steel parts in the vicinity of either of the test coils, because this can upset the machine balance.

Figure A.7. Conveyer coil from chute end showing aligment device (*By courtesy of Messrs. Rolls-Royce Ltd., Small Engine Division*)

(2) For conveyer testing, keep the successive components passing through the coil spaced at least one foot (300 mm) apart.

(3) Although the testing procedure will automatically de-magnetize the components, do not test components with strong residual fields resulting from prior machining on magnetic chucks or from magnetic particle testing.

(4) Do not wear a wrist watch in the vicinity of the test coils because they may magnetize the watch.

INDEX